Praise for *When Animals Rescue*

"If you thought that only humans were capable of virtuous behavior, think again! *When Animals Rescue* will open your eyes to the breadth and depth of nonhuman kindness. Belinda Recio weaves in fascinating facts with astonishing stories about all kinds of animals—large and small, domesticated and wild—acting in compassionate ways. We need this book—not just to learn more about how other animals demonstrate kindness, but also to inspire more of this behavior in our own species!"

—Marc Bekoff, PhD, author of *The Animals' Agenda,*
Canine Confidential, and *Unleashing Your Dog*

"Uplifting and life-affirming, *When Animals Rescue* will save your day! These true stories about animal heroes are a tonic for tough times. I loved it."

—Sy Montgomery, author of *How to Be a Good Creature:*
A Memoir in Thirteen Animals and *The Soul of an Octopus:*
A Surprising Exploration into the Wonder of Consciousness

"Social animals evolved to assist each other. This lovely book documents a great variety of rescues and helping, sometimes across species boundaries. There is really nothing to be ashamed of if someone calls you an animal!"

—Frans de Waal, author of *Mama's Last Hug: Animal Emotions*
and What They Tell Us about Ourselves

Praise for *Inside Animal Hearts and Minds*

"Belinda Recio has written a fascinating account of animal emotions and animal intelligence. She makes the stories she tells available to a wide audience, and her examples are fair, friendly, and charming. I cannot imagine any animal lover not finding this a wonderful book!"

—Jeffrey Moussaieff Masson, bestselling author of nine books on the emotional life of animals, including *Beasts: What Animals Can Teach Us About the Origins of Good and Evil*

"Belinda Recio has assembled the best true stories of animals' achievements in art, emotion, and reasoning all in one place. Everyone who reads it will come away feeling much closer to the other species we share this planet with, and be amazed by the sophisticated things they are capable of. . . . A provocative and beautiful book."

—David Rothenberg, author of *Survival of the Beautiful* and *Why Birds Sing*

YOUR
INNER
ZOO

YOUR INNER ZOO

A GUIDE TO THE MEANING OF ANIMALS AND THE INSIGHTS THEY OFFER US

BELINDA RECIO

Skyhorse Publishing

Skyhorse Publishing books may be purchased in bulk at special discounts for sales promotion, corporate gifts, fund-raising, or educational purposes. Special editions can also be created to specifications. For details, contact the Special Sales Department, Skyhorse Publishing, 307 West 36th Street, 11th Floor, New York, NY 10018 or info@ skyhorsepublishing.com.

Skyhorse® and Skyhorse Publishing® are registered trademarks of Skyhorse Publishing, Inc.®, a Delaware corporation.

Visit our website at www.skyhorsepublishing.com.

10 9 8 7 6 5 4 3 2 1

Library of Congress Cataloging-in-Publication Data is available on file.

Cover design by Belinda Recio and Marta Knudsen
Illustrations: Belinda Recio

Print ISBN: 978-1-5107-5794-3
Ebook ISBN: 978-1-5107-6703-4

Printed in China

In memory of Miklos Antal Radics,
who understood the gifts of animals, and gave them the gift of kindness.

To my husband, Ed,
to River, who found his way to us just when we needed him most,
and to all the animals, for all you give to us.

CONTENTS

In part I am trying to tell them something: a message about how they [animals] register in the human imagination, in our lore and fantasy, in our symbol systems, even what our zoology says about them. Like a report to them about how they are perceived.

—James Hillman, *Dream Animals*

The animals—and perhaps the plants—within us are like the beings of a larger and older reality. They exist within us in much the same sense that our parents and our ancestors are in us, not as ghosts but as shared form, a continuum of which we are only the present expression.

—Paul Shepherd, *The Others: How Animals Made Us Human*

ACKNOWLEDGMENTS

I am indebted to the scholars and scientists in the fields of animal studies, ethology, ethnozoology, ecopsychology, and other disciplines who are helping us to better understand the human-animal relationship and the roles that animals have played in shaping the human mind and culture.

Thank you to Skyhorse Publishing for supporting my work, and to my editor, Jason Katzman.

A heartfelt thank you to my friends and family, who support me in so many ways, and especially to my mother, Maria Radics, for her unwavering encouragement. Special thanks to Eileen London, for joining me, years ago, on the journey that eventually led me to this book.

I am immensely grateful to the following people for their help and generosity of heart: To my sister, Marta Knudsen, for her digital art expertise and assistance with the illustrations and cover art; to Mark Thayer, for his digital image editing and proofing; and especially to Joan Parisi Wilcox, for her insightful editorial contributions and steadfast support, not just with this book, but throughout so many projects over the years.

Finally, I extend a gratitude beyond words to my husband, Ed Blomquist, for all the ways he supported me throughout this project. I truly couldn't have done it without you.

AUTHOR'S NOTE

PRONOUNS AND TERMINOLOGY

Unfortunately, the English language does not provide a nonobjective gender-neutral pronoun to use when referring to animals. I do not use "it" because I do not want to objectify animals. Instead, I rely on gendered pronouns—both female and male—when referring to animals in the singular. In each chapter I assigned a pronoun to the featured animal and used that pronoun throughout, unless referencing a specific animal character in a folktale or myth, or an actual known animal in a scientific study. I wanted to stay true to the original source material, so please excuse any passing confusion in these instances.

Regarding terminology: although humans are animals, for stylistic reasons I use the term "animal" instead of "other animals" or "nonhuman animals."

ABOUT THE MATERIAL INCLUDED IN THIS BOOK

I have made every effort to ensure that the scientific information in this book was correct at press time, but it's possible that some statements might become inaccurate due to new discoveries, so please excuse any errors.

As for the cultural material that I included in this book, my primary goal was to present portraits of the animals that showed the connection between them and the ideas they have inspired in the human imagination. This objective, along with personal preferences and page count limits, shaped my choices. My hope is that I have captured the essence of these animals—how they appear in nature, as well as in our hearts and minds. That said, some readers may wish that I had included a particular folktale, symbolic aspect, or biological trait that I left out. Or they might draw different conclusions from the material than I did. If so, it means that the animals have nonetheless inspired ideas—even if different from my own—and worked their magic.

Due to limitations of space and similarities between certain species, I paired some animals in the same chapter—such as rabbit and hare, for instance—when I felt it made sense to do so.

For those interested in exploring further, please see the Selected Bibliography at the end of this book.

INTRODUCTION

> To reconnect to the animal, we must become aware of the animal in the psyche, the animal psyche, the animal in things, the animal in art, in world, in poems, in dreams, the animal that lies between us and the other.
>
> —Russell Lockhart, *Psyche Speaks: A Jungian Approach to Self and World*

Since prehistoric times, animals have held a special place in the human imagination because they have abilities that we lack, yet admire. Some can fly in the air, others can breathe underwater, and many can see, hear, smell, or otherwise sense the world in ways that are beyond our perception. Animals often know when the weather is going to change and which plants are poisonous. They know how to hunt, heal themselves, and build shelters. Early humans wanted to know what the animals knew, so they spent time watching and learning from them. They mimicked them in song and dance, wove them into mythologies, adorned themselves in fur and feathers, and painted and sculpted their images. Over time, observing animals helped humans to better understand the world and inspired the development of fundamental concepts. In this way, animals helped to shape the human mind as we evolved as a species.

The idea that the human mind developed through our relationship with animals has been explored by anthropologists, ecologists, philosophers, and psychoanalysts. The anthropologist Claude Levi-Strauss recognized the importance of animals to our understanding of ourselves and the world when he famously stated that "animals are good for thinking," meaning that they evoke ideas. Wallace Stevens, the twentieth-century poet known for his exploration of the relationship between the material world and the imagination, expressed a similar concept when he called animals "the first idea." Philosopher Gaston Bachelard believed that the need to "animalize," or view the world through an animal lens, is "at the origins of the imagination." Ecologist Paul Shepard suggested that animals awakened the human impulse to symbolize and engage in categorical thinking. Finally, archetypal psychologist James Hillman mused that animals may have been "the first psychoanalysts" because of their ability to make us aware of ourselves.

Consider, for instance, how early humans watched animals and developed concepts from their behavior. They saw *persistence* in a badger's unrelenting digging for roots, *indecision* in a butterfly's flitting from flower to flower, and *foresight* in a squirrel's caching of nuts for future consumption. These observations became ideas that people use to better understand themselves by inviting comparisons between humans and animals. Someone might "badger" his friends by engaging in repetitive, persistent behavior. A person who flits from one friendship to another might be considered a "social butterfly." Consistently saving for retirement might be described as "squirreling away" one's money.

It has always been easy to look for and recognize aspects of ourselves in animals—as opposed to other people—because humans tend to perceive characteristics, habits, abilities, and other qualities as more concentrated in animals. This is why our encounters with and observations of animals gave rise to animal symbols and idioms, such as those just discussed; folktales, such as that of the wily fox, who consistently outsmarts his opponents; cultural traditions, such as totemism, in which animals are used as emblems of identity; and shamanism, in which animals are believed to provide guidance and protection.

In *Your Inner Zoo*, I present many of the natural and cultural ideas associated with animals, and invite you to explore yourself and your life through a zoological lens. In doing so, you will hopefully develop what ecologist Paul Shepherd called a "zoology of the self"—a sense of how the other animals teach us and live within us as ideas and insights, intuitions and instincts.

As you explore your own inner zoo, may the animals herein teach, guide, and inspire you. In return, I hope that you will be inspired to help protect them and the places they live, as these creatures—along with all the rest of nature—hold the soul and wisdom of the earth, of which humanity is just a small part.

ABOUT THIS BOOK

HOW *YOUR INNER ZOO* IS ORGANIZED

Each chapter is focused on one animal and contains three sections: Keywords, Nature and Culture, and Insights.

Keywords

You can use the keywords listed at the start of each chapter to get an overview of the primary ideas that the animal has inspired in the human imagination. The keywords for all animals are also listed in the Keyword Index at the end of the book, and each keyword connects with all the animals associated with that particular theme. This cross-indexing enables you to explore any given theme across multiple animals. For example, the keyword "Transformation" is associated with the butterfly, frog, and seal, as well as other animals—and all of these animals are listed under "Transformation."

Nature and Culture

The Nature and Culture section explores the animal through his appearance, behavior, and ecology, as well as the ways in which the animal has been represented in human culture—through symbolism, idioms, metaphors, myths, folk beliefs, and the arts. This section is intended to help you connect with the animal by reminding you of the aspects of the animal that gave rise to his symbolic associations and cultural traditions.

Insights

Drawing from the Nature and Culture section, and introducing new ideas as well, the Insights section presents creative suggestions for how you might look at yourself and your life through the lens of the animal. This section relies on creative interpretation of the animal's natural and cultural history to explore psychological ideas associated with the animal. For example, the Insights section in the "Bear" chapter suggests that the bear's tendency to live off of stored fat during hibernation serves as a metaphor for our own ability to rely on our accumulated inner resources.

Combining Nature and Culture with Insights

Insights can be drawn from the information in the Nature and Culture section, just as additional biological and cultural information is sometimes revealed in the Insights section. To get the most out of the book, read both sections and let them play off of each other in your imagination. For example, a detail about an animal's ecological niche or her role in a creation myth might trigger insights that are not specifically highlighted in the Insights section.

WAYS TO USE *YOUR INNER ZOO*
Know Your Animals, Know Yourself

Animals can enlighten us about ourselves by inviting us to identify traits we share—or wish we shared—with other animals. Character traits tend to be more recognizable in animals, making such comparisons and contrasts relatively easy. By recognizing the myriad animal traits within ourselves—our inner zoo—we can develop a zoology of the self and deepen our connection with our kindred creatures.

Approach Animals as Teachers and Guides

You can use this book to approach animals as potential teachers—as sources of concepts and insights. The unique aspects of each animal—such as their abilities, behaviors, talents, and adaptations—suggest ideas and questions that can broaden your perspective or provide guidance or direction.

There have been many cultures throughout history who related to animals as teachers, guides, guardians, soul companions, and kin. You might have heard the terms *totem animals* and *spirit animals* to describe these special relationships. Understanding where these terms come from—and what they actually mean—will help you to forge your own authentic symbolic and spiritual relationships with animals, while simultaneously cultivating a sensitivity to the cultures whose traditions gave rise to these terms.

Let's start with the word *totem*, which is derived from Ojibwe—an Algonquian language spoken by a group of native peoples in the United States and Canada—and loosely translates as "clan" or "kin." Anthropologists use the term to describe a cultural system in which an animal, plant, or other natural element serves as an emblem, symbol, or source of spiritual kinship for a family, clan, or tribe. In popular usage, "totem animal" usually refers to an animal that you identify with or one who has traits that you desire or might believe you possess. Although the popular definition is vaguely similar to the actual meaning of the term, it is not the same.

As for *spirit animal*, an anthropologist might use the term to describe a person's spiritual connection with an animal. For example, the Oglala Sioux of the northern Great Plains have a rite of passage known as a "vision quest," during which a spirit—often in the form of an animal—might appear and impart strengths, skills, or spiritual insight to the seeker. Another example of what might be called a "spirit animal" comes from the Guatemalan Quiche culture, in which every child is born with a *nahual,* a protective animal spirit who accompanies the child throughout his life.

In contrast, the popular definition of spirit animal is an animal with whom one has a special affinity or whose characteristics a person embodies. More recently, the term has become an internet meme, in which it is used as a metaphor for someone or something a person relates to or admires. Again, although certain aspects of the popular definition of spirit animal are loosely connected to the anthropological term, it is not the same.

As such, you may want to avoid using these terms in the popular sense, because they perpetuate oversimplification and misconceptions of the indigenous traditions with which they are associated. By obscuring the original meanings of these terms with the ones in popular culture, we risk forgetting what the terms really mean and, consequently, could lose an important part of humanity's cultural heritage. Further, we risk offending those who belong to the cultures that practice these spiritual traditions.

That said, animals and all they have to teach us—by way of who they are and how they live, as well as the ideas and symbols they have inspired—do not belong to any *one* human culture. Humankind has related to and learned from animals since our species first evolved, and the opportunity to continue to do so is available to all of us. So feel free to approach animals as teachers, guides, metaphors, and symbols, but do so while being respectful of other cultures.

Recognize Animals as Signs or Messages

To consider the possibility that animal encounters might hold meaning, we need to briefly depart from the world of animals and dive into the ideas of synchronicity and an interconnected universe. Have you ever experienced random events or encounters that felt like a sign, or held meaning for you? If so, you are not alone—many people have had this kind of experience. In fact, in the early 1900s, analytical psychologist

Carl Jung believed that these kinds of meaningful coincidences happened far too frequently to be mere chance. For this reason, he introduced the term "synchronicity" to describe meaningful coincidences in which something other than chance appears to be involved.

Others also have considered the possibility that unrelated events actually might be linked. In 1964, physicist John Stewart Bell proposed what became known as Bell's Theorem, which asserts that everything in the universe is connected. A couple of decades later, quantum physicist David Bohm developed his concept of Implicate and Explicate Order, which proposes that everything in the universe affects everything else because they are all part of the same whole. Both of these theories allow for the possibility of meaningful coincidences.

But long before Jung, Bell, Bohm, and other modern Western thinkers developed such theories, many indigenous peoples around the world believed in the interconnectedness of all things and sometimes saw events in nature—such as certain kinds of animal encounters—as signs and messages sent from the spirit world.

If you are open to the possibility that meaningful coincidences exist, think about how this might apply to animal encounters. For example, say that within one week you repeatedly notice a soaring hawk, see a hawk bumper sticker, and receive a gift of a hawk pendant. Or, perhaps a hawk appears to you in a dream, meditation, prayer, or during creative inspiration. In all of these instances—whether in the physical or spiritual worlds—you might feel your attention was being drawn to hawks for a reason.

If an encounter with an animal feels like more than just coincidence—more like a synchronistic event—then use this book to look up the animal you encountered and consider the possible meanings. A hawk can see a rabbit from hundreds of feet in the air by reading the terrain, which could symbolize the ability to see into situations and interpret patterns and events. So, perhaps a hawk encounter might be calling your attention to the need to take a sharper look at what is hidden in the landscape of your life—an opportunity, a relationship that isn't what it seems, or perhaps even a part of yourself that you are keeping in the dark.

Reconnect With Animals

Many environmentalists and philosophers theorize that people are innately "biophilic" (defined as the human affinity for other life forms) and crave a sense of kinship with other creatures. Learning how an animal's natural history gave rise to myths, symbolism, and language re-enchants the human-animal relationship, which can then deepen our sense of belonging to the more-than-human world.

Explore Animal Symbolism

Your Inner Zoo can be used as an animal symbolism dictionary to explore the myriad meanings of animals in art, literature, mythology, and spiritual traditions.

Explore the Meaning of Animals in Dreams

Shamans, psychologists, and others have long believed that animals can appear in our dreams as messengers, teachers, or healers who are trying to tell us something about ourselves, our lives, or even the world. You can use *Your Inner Zoo* to explore the potential meanings of animals in dreams. Upon waking, record your dream with as much detail as possible. Then read about the animals who appeared in your dream, consider the context in which they appeared, and see what resonates for you based on personal associations and the current circumstances of your life.

Generate Ideas for Creative Projects

Looking for inspiration for a painting or a metaphor for a poem? *Your Inner Zoo* is a rich source of ideas and imagery for creative projects focused on animals or the human-animal relationship.

YOUR
INNER
ZOO

Community • Cooperation • Diligence • Foresight • Industry
Order • Prudence • Selflessness • Tenacity

ANT

THE ANT IN NATURE AND CULTURE

"The ants go marching," as the children's song reminds us. But these abundant social insects—with thirteen thousand species and roughly one hundred trillion individuals—do so much more than merely march. They work together to move mountains of dirt, systematically arranging tiny specks of sand into complex underground cities. They forage in trees, linking their bodies together to build bridges that span branches so they can cross a gap. During floods, ants assemble themselves into living rafts—a water-repellant lattice comprised of as many as one hundred thousand individuals—so that their colony can float for weeks and survive. Even when we trap them in sand-filled transparent boxes so our children can watch them, the ants still march on—burrowing, building tunnels, stockpiling food. They seem to be always moving, always working, and nearly always doing so together. No wonder so many cultures perceive ants as symbols of industry and cooperation.

But ants are more than just hard working team players. During times of plenty, they stockpile food for leaner days to come. Since ancient times, many cultures have taken note of ants' foresight. Within the Judeo-Christian tradition, ants represent prudence and virtue because they have the wisdom and discipline to plan ahead, which is why a biblical proverb admonishes us, "Go to the Ant, you sluggard; consider its ways and be wise!"

The Hopi tribe of North America also recognized the foresight that ants demonstrate. In their legend of the *Anu Sinom*, (Hopi for "Ant People"), the Hopi survived two world-destroying events thanks to the guidance and generosity of the Ant People. During each of these cataclysmic events, the industrious Ant People led the Hopi into underground caves, where they found shelter and food. When food eventually ran short, the Ant People shared theirs with the Hopi and taught them how to store food for the future. According to one version of the legend, the reason the Ant People have such slim waists is because they went hungry in order to feed the Hopi.

Ants work tirelessly, seem to never give up, and accomplish incredible feats, demonstrating the power of perseverance.

Like some human cultures, ant societies are rigidly structured. Their colonies operate as a caste system in which each ant is born into the role it will play for life: queens (reproductive females) or workers (reproductive males and non-reproductive females). There is no opportunity to break out of these roles and no social mobility. Societal order prevails. Because of their structured cities, hierarchical caste system, and strictly designated behaviors, ants exemplify order.

Ant societies share even more similarities with human culture. They are tool-users: using leaves, grains of sand, and other objects as containers for carrying honey and fruit pulp. They are accomplished architects who build elaborate multi-chambered dwellings connected by networks of horizontal tunnels and vertical shafts. They practice agriculture and animal husbandry. Leaf-cutter ants farm fungi, which they feed their colonies, and herder ants capture, herd, and "milk" aphids for a sugary food substance called "honeydew." Ants are among the very few animals—other than humans—known to capture, enslave, and domesticate another species.

Ants engage in warfare and employ a variety of military strategies, such as piling up pebbles at the entrance of a nest to prevent enemies from entering. One species of ant native to Borneo even includes suicide bombing in its arsenal of tactics. If the ant detects an intruder in his territory, he blows himself up by squeezing himself to death. The self-inflicted lethal compression sprays toxic liquid from a reservoir in the ant's body onto the enemy. The toxin kills the intruder before he can return to his army and report the location of the bomber's nest.

Yet another similarity between ants and humans is the habit of colonizing and invading. Ants will find their way into any space: natural structures such as plant stems and tree trunks, and even the most impenetrable of manmade structures. If there is an opening—no matter how small—ants will find it, force themselves in, and, if left unchecked, will destroy foundations and build massive colonies between walls. On a psychological

Because they stockpile food for leaner days to come, ants symbolize foresight and prudence in cultures around the world.

level, when ants creep into our psyches, they embody the very idea of infestation. This is why the sensation of insects crawling across or underneath our skin is called "formication," a term derived from the Latin word for ant, *formica*. Similarly, when we have "ants in our pants" we are so overcome with excitement or worry that it feels like an infestation.

But in other ways, ants are vastly different from us. They subordinate the individual to the colony, with each ant acting, in a sense, like a single neuron in a brain. Their very anatomy reflects the priority of the collective: they have two stomachs—one holds food for the ant and the other holds food to be shared with the colony. For all these reasons, we see ants as the many that act as one, the swarm that self-organizes, the otherworldly collective. No wonder the extraterrestrial societies in science fiction are often modeled on ants and other social insects. In these stories, the distinctly human perspective prevails: individual free will triumphs over the hive mind.

INSIGHTS FROM THE ANT

The ant's selflessness asks us to consider the extent to which we work together for the greater good, as opposed to working solely for ourselves. Ants remind us that sometimes we need to let go of our own personal ambitions and work cooperatively toward something that benefits our entire community. On the other hand, with their hive minds and rigid caste system, ants also can serve as a warning not to subordinate ourselves to the point where we blindly follow the prescriptions of society. Sometimes we need to exercise our free will and question the prevailing ethos so that we can correct course.

Like other animals who store food for the future, ants represent the value of forethought. In Aesop's fable "The Ants and the Grasshopper," a hungry grasshopper begs a group of ants for food. When the ants ask him why he didn't store food for leaner times, the grasshopper

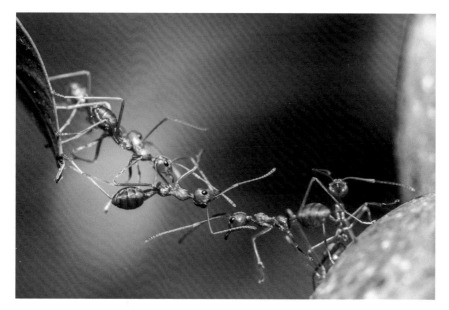

Reflecting the ant's inherently cooperative nature, army ants link their bodies together to build bridges between branches so that their comrades can march across.

explained that he was busy playing music. The ants shrug in disgust and tell him to "go dance." In this way, in a chorus of a trillion tiny voices, ants symbolically advise us to work hard and plan ahead, and there is no doubt that such prudence pays off. And yet, the world needs music, too, and herein lies another lesson that ants offer us by the negation of their example: Don't forget to play. Be a grasshopper . . . at least some of the time.

Finally, ants help us to remember that we are often stronger and more capable than we think. Ants work tirelessly—often against great odds—seem to never give up, and accomplish incredible feats. When we watch them carrying objects a thousand times heavier than themselves, we remember that what seems impossible sometimes isn't if we just try hard enough. Lyricist Sammy Cahn captured the tenacity of ants in his 1959 song "High Hopes," in which he describes an ant undertaking the impossible task of moving a rubber tree plant. What motivates the miniscule ant to try to move a massive rubber tree plant is nothing more—or less—than the power of diligence and perseverance. Perhaps this is the most valuable lesson these diminutive creatures offer us: think big and never give up.

The ant's tenacity can inspire us to think big.

**Dexterity • Immaturity • Instinct • Mimicry • Mischief
Playfulness • Politics • Shadow • Sociability • Trickster**

APE AND MONKEY

THE APE AND MONKEY IN NATURE AND CULTURE

When we observe most animals, we are often able to find aspects of their behavior that remind us of ourselves. But when we look at apes, monkeys, and other primates, it's difficult to find attributes that *don't* evoke comparisons with ourselves. Not only are we anatomically similar to our primate cousins, but we share other characteristics with them as well. To varying degrees, humans, apes, and monkeys are all self-aware, socially complex creatures who fight with one another and wage wars, comfort one another and reconcile, use tools, problem solve, and enjoy the dexterity provided by opposable thumbs and grasping hands.

That said, humans have long rejected equivalencies between people and other primates, seeing them as representing an earlier stage of human development, both physically and intellectually. Historically, humans have painted other primates as impulsive savages, which is why people who lose control of their emotions are sometimes described as "going ape." Similarly, the noisy chatter and frenetic behavior of monkeys inspired the Buddhist term "monkey mind" to refer to the unfocused, restless state of the unenlightened human mind.

However, there are many primal commonalities we share with our fellow primates. Like apes and monkeys, we fight over food, territory,

Like humans, apes and monkeys value their relationships and spend time building social networks and forging alliances, exhibiting a political savvy that is typical of primates.

and mates. Like them, we feel greed, lust, and aggression. Behaviors and emotions like these often are difficult for us to acknowledge in ourselves, so we relegate them to our shadow side while simultaneously projecting them onto our closest animal relatives.

And yet, we acknowledge that apes and monkeys also embody the admirable qualities of cleverness, playfulness, and humor. Often the very embodiment of mischief, they remind us of the urge we sometimes have to pull tails, steal a bite of someone's food, and otherwise "monkey around." Most primates—especially adolescents—are comical pranksters who like to test one another's limits. Like humans, they even seem to enjoy slapstick humor. Captive apes are known to laugh at gullible behavior in others, play tricks on their keepers, and steal things from visitors just for the fun of it, which is why they often are portrayed in myths and folktales as tricksters.

In the Chinese story, "Journey to the West," the character Monkey King—or *Sun Wukong* in Chinese—is a trickster god who in his youth violated many rules. He broke into heaven, drank the celestial wine, and erased his name from the book of the dead, thereby achieving immortality. Eventually his luck ran out and he was imprisoned for five hundred years, until Kwan Yin, the goddess of compassion, rescued him and gave him a chance to redeem himself by guarding a monk on a long and dangerous journey. Monkey King faithfully looked after the monk, and along the way learned from his teachings. He eventually achieved enlightenment.

Every animal has a talent that stands out so much that we associate it with them, such as the industriousness of a beaver or the memory of an elephant. When it comes to apes and monkeys, two skills stand out. The first is political savvy. These primates possess an impressive social aptitude that they use to form alliances, build coalitions, and engage

Historically, humans have painted other primates as impulsive savages, but many apes have gentle natures, avoid conflict, and possess a wisdom all their own.

in other political maneuvering. Even the primate talent for mimicry—which gave us the idiomatic expressions "monkey see, monkey do" and "to ape"—often plays a role in relationship building.

The second gift that nature gave apes and monkeys is their grasping ability. Whether climbing trees, using tools, grooming one another, or learning human sign language, these animals—like their human cousins—are capable of performing amazing feats with their hands, which is why we also associate them with dexterity.

Apes and monkeys remind us of our evolutionary origins and animal natures—aspects of our humanity that many of us deny.

INSIGHTS FROM THE APE AND MONKEY

Our fellow primates help us remember that we all have at least a little bit of the monkey's spirit. This monkey within can get us into trouble with her defiance and uncontrolled instincts, or she can help us to connect with our playful and comical sides. Mostly, she can help us remember not to take ourselves too seriously and to heed the warning that "pride goeth before a fall." Despite our high opinions of ourselves, we are clearly capable of the same foolish and immoral behaviors that we associate with monkeys and apes. But like the Buddhist maxim about the trio of monkeys who see, hear, and speak no evil, when it comes to our own base instincts, we often turn a blind eye, choosing to not see, hear, or speak of our own moral failings.

As masters of mimicry, monkeys also call our attention to the ease with which we copy and spread behaviors. They ask us to consider if we are too vulnerable to social trends that we follow in order to feel accepted. Instead of "monkey see, monkey do," there are times when "monkey see, monkey think" would serve us better.

Like humans, apes and monkeys have many wonderful qualities. They demonstrate empathy, console one another, have a sense of fairness, and often lend a helping hand. On the other hand, they also commit murder, wage war, and deceive and manipulate one another. Sound familiar?

Our fellow primates teach us that we are complicated works in progress. Imagining ourselves to be somewhere between animals and angels requires a balancing act in which we do our best to keep our primal impulses under control, while nurturing our higher selves. Like the Monkey King, we tend to break the rules and too often try to take shortcuts . . . but perhaps, in the end, we, too, will turn out to be redeemable.

Our simian cousins hold up a mirror for us. As we gaze into the eyes of the naked ape that is our own reflection, we can either acknowledge that we have many admirable and not so admirable apelike qualities, or deny our membership in the family of primates. Apes and monkeys therefore offer us a choice: stay in the dark and suffer the consequences of not knowing who we really are, or shine a light on our own shadows and deepen our self-awareness.

**Confidence • Cooperation • Courage • Determination
Domesticity • Endurance • Evasion • Shape-shifting • Tenacity**

BADGER

THE BADGER IN NATURE AND CULTURE

"Honey badger don't care" claims the narrator, known as Randall, in the viral video "The Crazy Nastyass Honey Badger." After watching a honey badger munching on a cobra (while surviving the snake's venomous bite), scooping out honey bee larvae from an active hive (while enduring sting after sting), and chasing off jackals twice his size, Randall's assessment seems fair. Nothing appears to deter the honey badger, who has earned the distinction of being called the world's most fearless animal. Although other species of badgers might not be quite as resilient as the honey badger, all badgers possess a fierceness of spirit that has made them symbols of tenacity, determination, and endurance.

Bold, independent, and assertive, badgers are quick to react and willing to fight if necessary. They will readily confront much larger animals, such as wolves and bears, in order to protect themselves and their families, which is why they also represent courage and confidence. But badgers appear to know their limits, too. When flight is a better strategy than fight—as is often the case when they are pursued by humans—they sometimes elude capture by running away so swiftly that they seem to vanish in plain sight.

The badger digs faster, longer, and with more intensity than any other animal on earth. He digs as if his life depends on it, and it does, because digging is how he survives. He digs for food, such as

The Arapaho, Sioux, and other Native American peoples believe that the badger knows which plants can be used as medicine.

earthworms, grubs, small mammals, and roots; he digs to create his sett, or underground den; and sometimes he digs to escape predators. The badger's digging ability is so extraordinary that the Navajo creation story features him as the animal who digs the passageway through which The People emerge into the world. On a less mythic and more practical scale, when the badger digs his sett, he excavates a series of tunnels that connect rooms used for different purposes, such as storing food, sleeping, and eliminating. His tidy domesticity inspired the badger characters of folktales and story who often are portrayed as provincial homebodies who simply want to live in peace.

Most of the time, the badger goes about his business under the cover of darkness, which might be why an old Apache story tells of a badger who once contained the world's darkness in a bag. (It was a Coyote who eventually released the darkness.) The badger's nocturnal ways, subterranean life, and uncanny ability to evade capture led to superstitions about the badger. For example, Japanese folklore portrays the clandestine badger as a *mujina*—a shape-shifting trickster.

When digging, the determined badger seems as if he's onto something, as if he knows what's underground even before he starts clawing at the earth. Perhaps this is why the badger came to be seen as a keeper of hidden secrets, especially those related to healing. The Arapaho, Sioux, and other Native American peoples believe that the badger has special knowledge of what lies beneath the earth—specifically plant roots and other natural healing materials. For the Sioux, the badger has such powerful healing capabilities that his medicine can be even stronger than the bear's. Across the Atlantic, the Picts—a group of Celtic-speaking peoples who once lived in northern Scotland—may have had similar beliefs about the wisdom of badgers, as their term for wise man or priest was *Brocan*, which contains the root word *broc*, meaning badger.

Despite his association with healing, the badger is also linked to death. Even today, encountering a badger is seen as a bad omen. Rural people in Great Britain have long believed that badgers perform ritualized burials, and there are reputable reports that support this belief. In his book, *A Country Chronicle*, naturalist Brian Vezey-Fitzgerald described a female badger surfacing from her sett late at night, raising her head to the sky, and vocalizing in a shrill howl. For the next two hours, she alternated between digging and descending to her den, and then once again, emerging and vocalizing, which eventually summoned a male badger. Her companion howled with her and then followed her into the den. A short time later, both emerged with the body of an older, dead badger, who they buried in the hole the female had dug earlier.

Several Native American tribes tell stories about badgers cavorting with coyotes, and science has proven these accounts to be true. Badgers and coyotes hunt together. They track small animals, such as prairie dogs and ground squirrels. When the prey is above ground, the coyote

When we find ourselves facing a daunting challenge, we need to conjure the fearlessness and tenacity of our "inner honey badger."

chases it; when it dives into a burrow, the badger takes over by digging and pursuing it underground. By sharing the workload and taking advantage of each other's specific hunting talents—coyotes run faster and have better eyesight; badgers dig faster and are better at sniffing out underground prey—both animals end up with a full belly more often than when they hunt on their own.

INSIGHTS FROM THE BADGER

The tenacious badger inspires us to stand our ground when we want something or feel threatened. There are times when, to get what we want or protect what we have, we need to be like the badger: we must react quickly, fighting fiercely and without hesitation. Badgers don't give up and they don't give in. They remind us to persevere and dig deeper within ourselves.

On the other hand, unrelenting badger behavior gave rise to the expression "to badger someone," which describes the act of incessant pestering. There are times when persistence can get us into trouble, or at the very least make us unpleasant to be around. Sometimes it's better to abandon the fight or change course by disappearing underground and re-emerging from a different position—another badger strategy. When all else fails, we can always take inspiration from the badger's alleged supernatural abilities and "shape-shift" into a more seductive form, one that might be more conducive to achieving our goals.

Another way the badger succeeds is through his collaboration with the coyote—a poignant reminder of the power of cooperation. If badgers and coyotes can reach across the species divide and work together to achieve a better outcome, then imagine what our species could do if we would simply reach across the political, religious, cultural, or ethnic gulfs that seem to so often separate us.

When we compare a badger and a bear, it seems clear who the stronger animal is. Yet, badgers willingly take on dangerous animals of all kinds, such as lions and cobras. They surprise us by defying expectations. These feisty animals highlight how assumptions—about ourselves and others—can limit us. The badger "doesn't care" about his opponent's advantages. He just goes for it, often winning his battles despite the odds, thereby showing us the power of bravado. Like the honey badger demonstrates, sometimes the only way to achieve a goal is to forge ahead regardless of the potential bites or brawls along the way. On the other hand, it's never a good idea to underestimate our opponents who might be conjuring their own inner badger.

**Ambiguity • Darkness • Death • Perception • Rebirth
Shape-shifting • Transition**

BAT

THE BAT IN NATURE AND CULTURE

In Aesop's fable, "The Birds, the Beasts, and the Bat," the birds and beasts declared war against each other, yet the bats would not decisively join either side. When the birds were winning, the bats claimed to be with them; when the beasts were winning, the bats placed their allegiance there. After the battle ended, the birds and beasts had a meeting and both sides agreed that the bats should be shunned for their refusal to clearly align themselves, and so banished them to the lonely edges of the night.

This fable is one of many that highlights how the ambiguity of bats has confounded humankind. As nocturnal mammals who fly on non-feathered wings, hang upside down, and use both sound and sight to find their way, bats defy easy categorization. They emerge at twilight and live betwixt mammal and bird, earth and sky. This liminality lent bats a supernatural air that led to associations with transitory states and realms—such as shape-shifting, the spirit world, and death—long before vampire bats were discovered by scientists in 1810. For example, an old Finnish folk belief holds that human souls rise from their bodies when sleeping, assume the form of bats, fly around the countryside, and then return to their human bodies in the morning.

In medieval Europe, bats became further linked to the darker aspects of the supernatural through their association with witches, demons, and even the devil himself, who was depicted with bat-like attributes. Later, in the early 1800s, when blood-sucking bats were discovered, they were named after the mythical blood-thirsty vampires from European folklore. Although only three of the more than thirteen hundred bat species in the world feed on blood, bats became so associated with vampires that, for many people, it is nearly impossible to think of one and not the other.

In addition to the supernatural, bats also have been linked to madness. When a cloud of bats emerges from a cave, erratically fluttering and flitting in the air, they create a feeling of chaos and disorder. This is the origin of the expression "bats in the belfry" and of the word "batty," which describes mental states ranging from scatterbrained to deranged.

Despite their darker associations, bats actually are shy, gentle creatures. They live in colonies—ranging in size from a hundred to several thousand individuals—and roost together, groom one another, develop long-term bonds, and devotedly tend to their young. Mother bats fly with their babies clinging to them, and when roosting sometimes cradle them in their wings. Even in the largest colonies—of a million bats or more—a mother bat can find her baby by exchanging specific calls with him. The bat's attentive mothering might be why some cultures associated them with mother goddesses.

Because bats live longer than other mammals of similar size, in Chinese culture they are a symbol of good luck. In fact, the Chinese character for bat, *fu*, sounds the same as the character that means good luck. In Chinese art, when five bats appear together—in what is known as the *wu fu* design—they represent the five blessings: health, wealth, peace, longevity, and a natural death. Bats also have been seen as symbols of rebirth because they emerge from darkness, as well as from womb-like caves—both of which are linked with the idea of gestation.

Bats navigate by producing very high-pitched sounds (outside the range of human hearing) and listening for the echoes, which their brains

We tend to project our fears onto that which is different and difficult to categorize, which is why some cultures made bats symbols of dark, demonic energies.

translate into map-like images of the environment. The echolocation of some species of fishing bats is so finely tuned that while flying above a pond they can detect the paper-thin fin of a minnow as it barely breaks the surface of the water. Because of their amazing ability to see in the dark and to use echolocation to sense their surroundings, bats are associated with the power to perceive things not present to the ordinary senses.

INSIGHTS FROM THE BAT

Bats beckon us into the dusky borderlands where definitions are blurred and the usual rules do not apply. While such ambiguity can be unnerving, it can also be liberating. The quintessentially betwixt bat invites us to think about how rigidly we define the world and ourselves. We don't always need to resolve ambiguity. When we can hold the tension that arises when two coupled ideas don't fit neatly together—such as *mammal* and *wings*—our thinking becomes less rigid and judgmental, and more flexible and tolerant. In contrast, when we struggle with this tension, we are likely to project our fears onto that which is different. Such projections played a part in bats becoming symbols of dark, demonic energies.

With their unique way of sensing the environment, bats call our attention to how we use our own senses. As humans, sight is our dominant sense and mediates much of how we perceive the world. But research has demonstrated that both blind and sighted humans can learn to use echolocation to "see" their environment, so we know that we have the potential to perceive in different modes. Bats invite us to consider how we prioritize some senses over others, and whether there are other ways to tune into our surroundings.

When we see bats hanging upside down, we instinctively want to turn our heads to get a better look at them. In doing so, we briefly see

Looking at bats, it's easy to understand why Aesop would write a fable about the ambiguous appearance of this flying mammal.

Bats beckon us into the dusky borderlands where definitions are blurred and the usual rules do not apply.

the world from their perspective. By inverting the usual order of things, bats help us to remember that sometimes we need to do a U-turn and look at things from an entirely different angle. Calmly hanging upside down, cloaked in their own wings, bats are the very picture of independent self-sufficiency. They whisper, "Do it your way."

Aggression • Creativity • Fortitude • Healing • Motherhood
Rebirth • Self-reliance • Shape-shifting • Strength • Wisdom

BEAR

THE BEAR IN NATURE AND CULTURE

In Alaska, the Koyukon people called him the "Dark Thing," whereas the Khanty of Central Asia referred to him as the "Old One of the Forest." To the Tlingit of the Pacific Northwest, he was the "One Going Around the Woods," for the Cree of Canada he was the "Chief's Son," while to the Mansi of northwestern Siberia he was the "Venerable One." Others simply addressed him with the exalted title, "Owner of the Earth." These are just a few of the countless euphemisms used by cultures throughout the northern hemisphere to refer to the bear. The peoples who shared their world with the bear considered it dangerous to refer directly to such a majestic and powerful creature—brown bears and grizzlies can grow to ten feet in height and weigh as much as fifteen-hundred pounds. They presumed it was safer—and more polite—to speak of the bear obliquely.

Back when the bear had an extensive range throughout Europe, Asia, and North America, those who encountered the great beast recognized his immense strength—a quality we still attribute to bears evidenced by numerous English idioms, including "to bear (something)," meaning to have the strength and fortitude to endure a situation or struggle. When the impressive and terrifying bear was angry, he was wrathful and unstoppable in his rage, which is why the bear is associated

Female bears are fiercely protective of their cubs, play and cuddle with them, and teach and discipline them, all of which gave rise to the metaphor, "mama bear."

with brute force and aggression. The bear's symbolic association with unchecked ferocity was used by a group of Viking warriors who, hoping to channel the irrepressible fury of a fighting bear, adorned themselves in bearskins when going into battle. These warriors were called *berserkers*—from *ber* for "bear" and *serkr* for "shirt"—and gave us the word "berserk," meaning frenzied and furious behavior.

Because of his strength and ferocity, people who shared their world with the bear endowed him with supernatural powers, such as being able to control other animals—especially the game animals on which they relied for survival. It was, therefore, up to the bear whether they starved or not. As a consequence, he was respected, regarded as sacred, and even worshipped. But the bear was more than just fiercely powerful to those who knew him—he also was uncannily humanlike. Like people, the bear can stand upright on the soles of his feet and can use his front paws like hands. He can gather fruits and nuts, fashion backscratchers from branches, and throw rocks, snowballs, and other objects. The bear even seems to share our moods: he hums when lumbering through the landscape, smiles when relaxing and playing, and gets gruff with mischievous cubs.

The bear's similarities to humans caused many cultures to see him as a relative, a mediator between the human and animal worlds, or a shape-shifter. For example, there are several versions of an old story, found from North America to Siberia, about a woman who married a bear and gave birth to the being from which all of humanity is descended. There also are many stories of bears as animal bridegrooms who transform into men, such as those found in fairy tales like "Beauty and the Beast" and "East of the Sun and West of the Moon."

Early peoples observed bears and noticed how they moved through the landscape and the seasons with purpose and intention. They saw how bears always find their way, which might be why, when we need to determine our position or situation relative to our surroundings or context, we say that we need to get "our bearings." Bears not only know how to navigate space, but they also seem to know about how the landscape changes throughout the seasons—not only when berries will be ripe and salmon running, but where to find them. Because the bear seems to remember everything and forget nothing—because he could "bear (it) in mind"—he also is seen as wise.

Having a good memory is important to bears because, as omnivores who hunt and forage, they need to remember where they can find the prey and plants that sustain them. Like humans, bears are extremely adaptable omnivores and can make a dinner out of just about anything, from fish and rodents to roots, tubers, grasses, and flowers. Bears don't just forage for plants when hungry—they also gather them for medicinal reasons. Biologists who study self-medicating behavior in animals (called Zoopharmacognosy) have observed bears using medicinal

The Inupiaq people of Alaska claim that they learned how to hunt seals and build igloos from the polar bear, suggesting that the lessons learned from animals helped early humans to survive.

plants in an apparent attempt to heal themselves. For example, brown bears chew the roots of osha—a perennial herb related to carrots—into a pulp, mix it with saliva, and make a paste they use to treat—and possibly even prevent—insect bites. Many Native American tribes knew about this behavior long before modern science observed it. Tribes such as the Apache, Navajo, Zuni, and others have historically treated a variety of ailments not only with osha, which they call "bear root," but also with other herbs that they learned about by watching bears. For this reason, many native tribes consider the bear to be the consummate herbalist and believe that their most powerful healers acquire their knowledge from the bear spirit.

Because of the fiercely protective way that mother bears defend their cubs, the bear also is linked to motherhood and the creative force. The mother bear's nurturing begins the moment she gives birth. When bear cubs are first born, they are tiny, blind, and have only a light covering of fur. The mother bear immediately begins to lick her cubs, continuing until they start moving. The mother bear's behavior (licking) and the apparent outcome (wriggling bear cubs) made such an impression on the human imagination that it gave rise to the expression "to lick into shape," meaning to carefully tend to someone or something to bring them or it to full potential.

Every winter, in order to survive colder temperatures and food scarcity, the bear goes into his den and enters a hibernation-like state of sleep, known as torpor. His breathing and heart rate decreases, his body temperature lowers, and his metabolism slows. During torpor, the bear lives off stored fat and is reliant only on his own body for nutrients, which links him to the idea of self-reliance. His springtime emergence from his den, in which he appears to have awakened from a death-like sleep, connect him to the concepts of rebirth and regeneration.

INSIGHTS FROM THE BEAR

As one of the animal kingdom's most flexible omnivores, the bear has much to teach us about adaptability—not just in diet, but in mindset as well. The bear can usually find something with which to feed or heal himself. This flexibility doesn't result merely in a full belly or a healed body, but also fosters a familiarity with seasonal patterns and locations. This bearish behavior can inspire us to be similarly flexible and open minded as we look to nourish or heal ourselves, both literally and metaphorically. When we broaden our perspectives, we are more likely to find alternative ways to address our needs. Furthermore, we are more likely to learn about *when* and *where* to look for what we need. In doing so, like the bear, we develop an invaluable knowledge base that we can rely upon throughout our lives.

The bear not only knows where and when to find food, he also knows when to retreat to his den to ride out a challenging time. Like the bear,

Many indigenous peoples perceived the bear as a relative, shape-shifter, or mediator between the human and animal worlds because of his humanlike behaviors and upright stance.

we sometimes feel a need to retreat from the world, particularly after periods of stress or upheaval. When we feel this ursine urge to carve out restorative time to "hibernate," we should think of it as the sensible impulse of our inner bear. We might consider withdrawing from some social activities to take stock of our lives, start a creative project, plan a trip, or plant seeds of thought that will hopefully germinate and come to fruition in the future. However, we also need to remember that bears come out of their dens once spring arrives. Spending too much time in isolation can deprive us of connection with and inspiration from the outside world. It's best to balance the urge to retreat for replenishment with the opportunity to be revitalized by all that the world has to offer.

For those of us with the drive to create—and the sometimes impatient wish for inspiration to come when it doesn't—the image of a mother bear emerging from her den with newborn cubs offers us a valuable reminder: creativity sometimes requires periods of dormancy, solitude, and quiet in order to bring forth something new. And the image of the mother bear "licking her cubs into shape" helps us to remember that bringing anything into being requires effort and dedication.

The counterpoint to the bear's link to creativity is his association with destructive forces, which can remind us to pay attention to our own overwhelming emotions and the ways they can devour us. When we feel outraged, do we deny the urge to express it, keeping our feelings hidden under a veneer of calmness? Or are we more inclined to act like bearskin-clad warriors and go berserk? Either way, consider the middle road, inspired by advice offered to those who encounter a bear while hiking: do not run, do not fight, but instead hold your ground. When we meet the fury of our inner bear, we can, metaphorically speaking, take the same approach: don't deny him, but don't indulge him. Instead, we can hold our ground until our inner bear moves on. In doing so, we build the endurance we need to better *bear* whatever we might encounter.

The bear's gift to us is this: he helps us to mediate between the human and animal worlds, and provides a mirror into our interior dual nature. We have two selves: an instinctive animal self and a cultured human self. The bear can teach us to accept both sides of ourselves and to learn when to let each self take the lead. Sometimes we need to be primal and instinctual and, without overthinking things, let our bodies lead us as we follow an impulse. At other times, our intellect and reason provide the right guidance and a safer path. If we are able to shape-shift between these aspects of the self, we enlarge our range of responses to the world.

Cooperation • Creativity • Diligence • Domesticity
Enthusiasm • Industry

BEAVER

THE BEAVER IN NATURE AND CULTURE

Imagine a beaver swimming through a pond and emerging from the water carrying an armful of mud in his forepaws. He then stands upright and, walking on his two back legs, climbs onto the roof of a domed shelter constructed of logs, sticks, and grasses. Using his hand-like paws, he applies the mud to the roof, packing it into place, and patting it down. The beaver then dives back into the water and returns later, carrying logs from small trees he cut down with his own teeth. He uses these to reinforce another part of the lodge. Given such behavior, it's not surprising that beavers are so often compared to humans. Standing upright, making structural adjustments to his home, a beaver looks more like a handyman than a rodent.

Beavers are one of the workaholics of the animal world. With teeth that never stop growing, they spend most of their time tirelessly gnawing on trees, branches, and twigs. They then haul their building materials—logs, branches, mud, grass—to their sites to construct lodges, dams, and canals. Beavers are so hard working that they can complete a lodge as large as a two-car garage in less than three weeks. Their incessant activity gave rise to the expressions "busy as a beaver" and "eager beaver," both used to describe someone who works hard and is always ready to take on a project. As a result of their work ethic and

Beavers offer us an image of a balanced life comprised of hard work, domesticity, and creature comforts.

communal building, beavers have long symbolized industry, diligence, and social cooperation.

Beavers live and work in colonies of about six individuals. All the members of a colony work together cooperatively when building and maintaining their lodges and dams. The beaver's architecture—the most complex of all mammals, except for humans—is similar to ours, not just in its construction methodology, but also in its floor plans. Beaver lodges have multiple chambers—one that serves as a nursery, another for feeding, and a separate quarters for sleeping. The lodge's multiple entrances and exits incorporate a unique security measure—they are underwater, which makes it difficult for most predators to reach the residents. However, beavers—with their webbed feet and rudder-like tails—are strong swimmers who can hold their breath for as long as fifteen minutes. Being able to spend so much time underwater enables them to not only use their secure entrances and exits, but to more efficiently build the submerged parts of their lodges.

Inside the beaver's lodge we find the picture of domesticity. A monogamous couple (beavers mate for life) and their young (sometimes from more than one litter) occupy the multi-chambered lodge. Family bonds are strong and stable, aggression is rare, and food is shared. Beavers plan ahead, so in the late fall they stockpile food in a larder for the long winter to come. Because of their devotion to family and good housekeeping habits, beavers are associated with domestic fidelity and creature comforts. Their prudence in planning ahead, as well as their industriousness, inspired a myth belonging to the Tlingit nation of the Pacific Northwest. The story tells of a porcupine who stole the provisions a beaver had stored for the winter. To retaliate, the beaver and his friends swam the porcupine to an island and left him there to perish. But once the waters around the island froze, the porcupine returned, gathered his people, and carried the beaver to the top of a tree and left him there to die. Of course, the beaver simply ate the tree from the top down and survived.

If a beaver wants to build a lodge and the water isn't deep enough for submerged entrances, he simply builds a dam to deepen the water. In doing so, the beaver doesn't merely accomplish his own goal of having less accessible entrances, but actually changes the entire environment. Beaver dams create habitat, prevent erosion, reduce floods, help stop the spread of wildfires, and even purify the water. Ecologists refer to beavers as a "keystone species," due to the fact that they play such an important role in maintaining the structure of their ecological community.

This "world-making" ability is highlighted in Native American creation stories, in which the beaver plays an important role in flooding the Earth or rebuilding it with mud after it was submerged. In other stories, the beaver was credited with the formation of specific bodies of

Most of us know at least a few "eager beavers"—individuals who dive into a project with enthusiasm and seemingly endless energy.

water, such as the Great Lakes. Science and myth differ in the details, but both recognize the beaver as a powerful creative force who shapes the landscape.

INSIGHTS FROM THE BEAVER

Because beavers are shapers of their environment, they demonstrate how local changes can have wide-ranging positive impacts. These industrious rodents show us that the right kind of creativity has the power to change the world for the better. If beavers can change the course of a river, reshape the terrain, and create environments rich in sustenance, we too can reshape our own lives to increase our well-being and bounty.

Consummate team players, beavers would be unable to accomplish their amazing engineering feats if they worked alone or uncooperatively, and so call our attention to the value of working together to achieve a common goal. Cooperating with others not only results in accomplishments that would be otherwise impossible, but also can yield benefits that increase our chances of survival. Even more impressive, the beavers' shared work is not finished when the initial goal is reached. They maintain their homes and dams by regularly inspecting them for leaks, and by using a wide variety of materials to make any required repairs. In doing so, beavers remind us that creating something is often just the start of an endeavor that requires ongoing diligence.

Finally, a beaver family tucked into their cozy lodge with plenty of food stashed away for winter highlights many of the elements important to a successful and satisfying life: the pleasures of family, the rewards of hard work and cooperation, and the wisdom of planning ahead.

Determination • Rebirth • Renewal
Self-development • Transformation

BEETLE

THE BEETLE IN NATURE AND CULTURE

In ancient Egypt, the scarab—a species of dung beetle—symbolized the lofty ideas of resurrection and self-generation. This may seem like powerful mojo to project on modest little beetles, but their unique behavior helps explain this symbolism. The male scarab shapes dung into a marble-sized sphere and rolls the ball along as he seeks out a female. The dung ball is his mating offering, which the female will use as a brooding ball (encasing her eggs in it) or as food. He pushes the tiny sphere across the ground from east to west, which reminded the ancient Egyptians of the daily solar cycle, in which the sun that "dies" every night at sunset is "resurrected" every morning at sunrise.

The scarab came to represent the idea of self-generation because, when the hidden eggs hatched and young beetles emerged from the dung ball, it seemed to the Egyptians—who didn't know about the eggs—that the beetle possessed the power to create himself. Believing that the scarab had special powers, they named him *kheper*, meaning "rising from, coming into being itself." This belief became personified in their pantheon as *Khepri*, a scarab-headed god of rebirth and self-creation. It also gave rise to the use of scarab amulets that symbolized rebirth, renewal, and self-development.

The scarab beetle is the embodiment of determination as he rolls his nuptial gift across the earth, guided by the light of the Milky Way.

Roughly five thousand years later, long after the Egyptian's belief about the scarab's solar symbolism faded into history, scientists discovered that this once-sacred beetle really does have a celestial connection. While traveling with his nuptial gift, the beetle climbs on top of his sphere and spins around. He does this when he encounters an obstacle as he pushes the sphere along or when the ball rolls away from him and he finally recovers it. Scientists had no idea why scarabs did this spinning dance. Then, through a series of experiments, they figured out that the beetle's dance atop his dung ball is for orientation: he is actually checking the location of the sun, moon, or Milky Way, using the position of the celestial object to set or reset a straight course. Then he is off again, traveling with his dung sphere (his only hope of winning a mate), and nothing seems to deter him—not difficult terrain, roll-aways, or even other males, who sometimes try to steal his sphere instead of making one of their own. For this reason, the resolute scarab also represents determination.

The scarab beetle is just one of three hundred and fifty thousand different species of beetles. One of the distinguishing features of all beetles is their front pair of wings—called the *elytra*—which are hard, protective cases that cover their softer, more vulnerable parts. Another feature that all beetles share is the process of metamorphosis. Every beetle begins life as an egg, lives for a time as a larva (also known as a grub), enters a pupal stage (during which he begins to take the shape of the adult beetle), and then emerges as a winged beetle. Through his remarkable metamorphosis, the beetle is linked to the idea of transformation.

INSIGHTS FROM THE BEETLE

The determined dung beetle is the picture of perseverance. Nothing stops him in his pursuit to deliver his nuptial gift and procreate. He may, like the rest of us, lose his way occasionally, but even that doesn't deter him. Instead, he takes the time to re-orient and get his bearings. In doing so, he reminds us that determination alone isn't enough. In our lives, like the beetle's, in order to arrive at a destination or accomplish a goal, we need to know where we are relative to our surroundings or situation. We need to have a reference point, whether it's the Milky Way or a guiding principle, and never lose sight of it.

As a result of the ancient Egyptian interpretation of the scarab's behavior, he is forever associated with the ideas of rebirth and self-creation. Thousands of years later, the scarab still symbolizes these ideas

The beetle's transformation from a soft-bodied grub to a hard-shelled, winged, and often dramatically-colored creature is one of nature's most dramatic metamorphoses.

and serves as a reminder that we are never completely finished with the process of becoming who we are. Like Khepri, we are always "rising from and coming into being." Consequently, to fully realize our potential—to achieve self-actualization—we need to find ways to nurture dormant parts of ourselves and make room for these aspects of self to come into being.

The beetle's metamorphosis offers us a metaphor for profound personal transformations. The start of such a transformation could be compared to the beetle's larval stage, during which he repeatedly sheds his exoskeleton. At this point, we might find ourselves changing the way we present ourselves to the world because our circumstances and priorities are shifting. Further into the transition, we enter the pupa stage. For the beetle, this is when his cells are literally rearranging themselves, turning him into an entirely new creature. For us, this phase is when we come to terms with whatever triggered the metamorphosis—such as having a child or changing careers—and we "rearrange" ourselves in order to become the person we now want or need to be. As challenging as life's transitions can be, the beetle shows us that we will get through them and emerge with our own new way of being.

The scarab also shines a light on how science and our understanding of the world are always transforming. The ancient Egyptians dubbed the dung beetle *kheper* because they believed he had the power to self-generate. Today, scientists refer to him as *Scarabaeus sacer*, and tell us that he looks up at the firmament and makes celestial maps in his mind. As he rolls his sphere under the starry skies, this little beetle helps us to remember that the magic in the world never vanishes, it just changes with our perspective.

Abundance • Fertility • Interconnectedness
Life Force • Strength

BISON

THE BISON IN NATURE AND CULTURE

In the classic folk song "Home on the Range," buffalo roam with deer and antelope in a pristine wilderness. The familiar song helped to make the American bison, more commonly known as the buffalo, one of the quintessential symbols of the American West. As many as sixty million bison once roamed on western grasslands. These animals were profoundly important to the Plains Indians, whose way of life was intimately interwoven with the bison who provided food, clothing, shelter, tools, and weapons. Bison are also an important part of the spiritual heritage of the Plains Indians, symbolizing the sacred interconnectedness of life, sustenance, and abundance.

Bison, buffalo, and cattle all belong to the same biological family, and share certain physical characteristics, such as a massive body, immense strength, herding behavior, cloven hooves, and horns. Cattle are the domesticated species that early peoples bred from an animal known as the auroch, the bull-like horned creature depicted in prehistoric cave paintings and invested with symbolic significance since the dawn of human consciousness. These animals played such an important role in the development of human civilization that *aleph*, the name of the first letter of the Hebrew alphabet (the letter A), means ox, and that letter's shape evolved from an ancient Egyptian hieroglyph of an ox's head.

For the North American Plains Indians, bison—perceived as sacred sources of sustenance—are seen as a manifestation of the divine and a reminder to practice gratitude.

Throughout history—and around the world—these bullish animals have symbolized strength, vitality, and fecundity. The Sanskrit words for bull and rain come from the same root word, illustrating the extent to which bulls were linked to the life force and fertility. Their heavy bodies have been associated with the Earth itself, as seen in Hindu cosmology in which the legs of a bull support the planet. Since being domesticated over ten thousand years ago, cattle have provided humans with hides (clothing and shelter material), food (meat and milk), and work (plowing, pulling carts and wagons, powering millstones, and other primitive machines). Given all they supplied to their owners, cattle quickly came to be associated with wealth.

In contrast with bulls, cows are associated with gentleness, docility, and nurturance. There are many stories of primordial bovines and cow goddesses around the world. In Norse mythology, for example, the meeting of fire and ice created Audhumla, a primordial cow, who nurtured the first beings. Within the North American Lakota tradition, the legend of White Buffalo Woman relates how a beautiful woman dressed in white brought the Lakota the sacred pipe, taught them how to pray with it, and told them of the sacredness of the buffalo. Upon leaving them, she turned into a white female buffalo calf and became known thereafter as White Buffalo Woman. This story connects the bison, who roamed by the millions across the plains, to abundance and manifestation through prayer.

INSIGHTS FROM THE BISON

From free-ranging bison to domesticated cattle, these majestic animals have been nourishing and working for humankind for millennia. They have long served our species to the profound detriment of their own. Because of all they have given us, bison (and their related cousins, cattle) can inspire us to practice gratitude for nature's bounty and to have respect for the sacred web of life in which all of us are bound.

As herd animals, bison remind us of the benefits of looking out for one another. When, to provide protection from predators, female bison form circles around their calves, and the males encircle the females, it provides a powerful image of the bonds of community. But herds can spread panic, too, and the fury of a stampede evokes comparisons with the dangers of an uncontrolled, unfocused group mind. Bison, therefore, can also serve as a cautionary reminder not to blindly adopt the behaviors of others or give into mob mentality.

Bison and other bovines are ruminants—animals with a unique digestive system that allows them to chew their food more than once, technically called rumination, but better known as chewing the cud. At some point in history, people likened the image of slow and aimless chewing with deliberative consideration, which is why ruminants can teach us to take the time to really *chew on* something, or carefully think things over.

The image of a charging bull evokes the idea of moving forward without hesitation or fear to meet challenges head on.

Powerful animals with impressive endurance, bison invite us to find our inner strength and *plow through* those times when we feel too depleted or discouraged to finish the task at hand. The bull, who can weigh up to two thousand pounds and stand six feet tall, is the perfect metaphor for something that seems invincible. This is why, when we finally summon the courage to confront a seemingly insurmountable problem head on, we might say that we are taking the bull by the horns.

Right before charging, a bull will paw the ground and kick up clouds of dust, giving his rival fair warning and a chance to back off. Despite their reputation for being stubborn, a rival bull will often walk away or offer submissive gestures to defuse a situation, rather than charging and locking horns. So the bull—ironically—can teach us not to be bullheaded. He shows us that there is more than one way to approach even the most bullish of obstacles.

Most of the time bison are peaceful, quiet animals, with the contemplative air typical of ruminants—animals who "chew the cud."

**Inconsistency • Indecision • Metamorphosis
Soul • Transformation**

BUTTERFLY

THE BUTERFLY IN NATURE AND CULTURE

One night, long ago in China, a philosopher named Chuang Tzu fell asleep and dreamt that he was a butterfly. When he awoke, he realized that he was not a butterfly, but simply Chuang Tzu *dreaming* he was a butterfly. He then paused, reconsidered, and asked himself, "Was I Chuang Tzu dreaming I was a butterfly or am I actually a butterfly now dreaming that I am Chuang Tzu?"

The meaning of Chuang Tzu's "Dream of the Butterfly" parable has long been debated. One common interpretation proposes that Chuang Tzu was asking how can we be sure of reality when both waking and dreaming seem so real? Another interpretation sees the story as reflecting the way the world is always changing from one form or state of being into another. This analysis makes the butterfly the ideal choice for the parable because no other animal so perfectly embodies the idea of transformation.

The butterfly starts life as an egg, hatches as an earthbound caterpillar, then entombs herself in a coffin-like chrysalis from which she finally emerges as a beautiful winged creature of sunlight and air. Her metamorphosis is one of the most profound in nature, and has made her a universal symbol of transformation and rebirth. Here's how it happens: When a butterfly is in her larval stage—what we call a caterpillar—she

The butterfly's epic metamorphosis epitomizes the idea of reinventing oneself.

eats and grows and goes through a series of molts in which she successively sheds her old skin. When it's time, the caterpillar stops eating, finds a spot to anchor herself, and molts into a chrysalis, which gradually hardens to form a protective shell. What happens next is startling. Inside the chrysalis, the caterpillar releases enzymes that dissolve her own tissue, essentially liquefying herself. Then, specialized groups of cells—called imaginal discs—begin to divide and multiply, slowly rebuilding the caterpillar as a butterfly. When the metamorphosis is complete, the butterfly breaks out of her chrysalis and flies away.

The butterfly's mysterious three-stage life cycle stirred the human imagination such that it became a metaphor for sacred change of all kinds. For example, in the Christian tradition, the caterpillar stage represents mortality, the chrysalis stage signifies the resurrection from the tomb, and the emergence of the butterfly evokes the miracle of the risen Christ.

In addition to being seen as a symbol of transformation, the airborne butterfly is also associated with the survival of the soul after physical death. The Australian Aborigines, as well as people of other cultures, believed that butterflies are the souls of the dead returning from the afterlife. The association between the butterfly and the soul is embodied in the Greek myth of Psyche, whose name means both soul and butterfly.

Like other creatures who fly, the butterfly has been seen as a messenger. The traditional beliefs of the Tohono Oodham tribe (of the American Southwest) hold that the butterfly can deliver prayers to the Great Spirit. They would capture a butterfly without harming her, whisper their prayers to her, and then set her free.

Another aspect of the butterfly that made an impression on people is the way she flies when feeding. Because she flits from flower to flower, alighting here and there, the butterfly came to represent indecision and inconsistency, as in the expression "social butterfly." This symbolic association is especially common in Japan, where the butterfly is linked with female vanity, unfaithful lovers, and Geishas. Similarly, the Spanish word *mariposa* means "butterfly"; however when used as slang, it means prostitute.

INSIGHTS FROM THE BUTTERFLY

The butterfly's seemingly fickle way of feeding can call our attention to a lack of focus and commitment in our lives—all too easy a trap to fall into during a time when so much of our lives are spent surfing the internet and ruling out prospective dates with a single swipe. If we constantly flit from one job, relationship, or interest to another, it's unlikely that we will find what we are looking for, let alone feel nourished or enriched. In order for life to be meaningful, we need to linger longer, pay attention, and dive deeper—at least on occasion.

One way that the butterfly takes us deeper is through her metamorphosis, which can be a powerful metaphor for the emergence of hidden

When butterflies developed the ability to metamorphose, they expanded their ecological niche, enabling them to go places they could never get to as caterpillars, reminding us that transformation brings new opportunities.

potential. The butterfly's transformation calls to mind the Zen saying, "No seed sees its flower." No caterpillar sees her butterfly, but a butterfly she will become, nonetheless. In this way, the butterfly reminds us of the latent possibilities within ourselves. We need to honor the butterfly within by paying attention to our callings, pursuing our passions, and being open to the kind of profound change that is sometimes required in order to more fully realize our capabilities.

Remember that the butterfly liquefies herself in order to turn from a fuzzy worm into a spectacular winged creature. She helps us to accept that big changes—those worthy of the word "transformation"—often require a metaphorical chrysalis stage. During this liminal time, when we are neither our old selves nor the new selves that we will become, we often need to let go of ideas, identities, or other aspects of our lives in order to truly change. Significant transformations of self nearly always require a certain amount of isolation, discomfort, and sometimes even suffering. But as the butterfly, by way of the Chinese philosopher Lao Tzu, points out: "What the caterpillar calls the end, the rest of the world calls a butterfly."

Biologists had a moment of poetic inspiration when they named the cells responsible for a butterfly's transformation "imaginal cells," evoking the idea of "imagining" profound change during times of disintegration.

**Endurance • Inner Resources • Journey • Obstinance
Service • Stamina**

CAMEL

THE CAMEL IN NATURE AND CULTURE

There is an Arab proverb that claims Allah shared ninety-nine of his names with humankind, but only shared his one-hundredth name with the camel. Perhaps this is why the camel wears what some have described as a calm smile of superiority on his face. Being the keeper of such a powerful secret also might be the reason why nothing seems to faze the self-possessed camel—not desert heat, sandstorms, nor a lifetime of long journeys during which he carries the heavy loads of others upon his back.

The camel is known for his ability to go for long periods—sometimes weeks—without water, as his hump stores fatty cells that can be metabolically converted to water. His thick coat and tough skin provide insulation against the extreme heat and wind. He is protected from sandstorms by nostrils that close and eyes that have a clear, shielding inner lid. His padded feet—comprised of two webbed toes that spread to prevent sinking in sand—can tolerate the high temperatures of the hot desert. As a result of all of these remarkable adaptations, the camel has long symbolized endurance and stamina.

Because the peoples of the Middle East long relied on the camel for travel—most notably the nomadic Bedouins of the desert—the camel came to be associated with journeys. Since first domesticating

Camels are passionate creatures, prone to a spectrum of strong emotions, and possessed of a soul soothed by music.

the resilient camel around five thousand years ago, humans have used them as a source of meat and milk, but also have been riding, racing, trading, and adorning them. As a result of their multi-faceted service to humanity, along with their minimal needs, camels have come to represent humility and the willingness to bear another's burden.

Like many domesticated animals, camels are quite intelligent, with cognitive abilities that have been compared to those of dogs; they can be taught to follow commands and have impressive memories. But even more noteworthy than their intelligence is their passionate nature. They respond to music, and so cameleers (camel handlers) often sing to their camels because they believe that a camel works harder and is happier when sung to. When it comes to the strongest emotions—love and hate—camels appear to feel both fiercely. A human who has won the trust and affection of a camel might be showered in kisses from his soft, velvety lips or gently nuzzled face to face as the camel's long lashes brush against her cheek. But a camel who has been wronged will not forget it. At some point, when a perfect opportunity comes along, he will retaliate and take his revenge—sometimes fatally.

Camels are hard to intimidate and can be stubborn and obstinate, preferring to do things their own way and sometimes refusing to carry a load. Among those who live and work with camels, there is a common folk belief that camels only accept burdens that they can carry. Because they appear to know their own limits, they are associated with discernment; but their obstinate independence has also linked them to stubbornness.

INSIGHTS FROM THE CAMEL

As an animal who can survive and persevere in harsh conditions, the camel invites us to think about how we endure difficult circumstances

Known for his ability to endure harsh conditions that would test the best of us, the camel is associated with the idea of the pilgrimage—a transformative journey made for spiritual reasons.

in our own lives. Do we have sufficient inner reserves on which we can rely during times of scarcity? Can we "take the heat" by finding ways to insulate ourselves from stressful situations? Or do we run for the shade whenever things gets too hot? Do we know where—or who—our oases are when journeys are arduous or times are tough?

The discerning camel also prompts us to consider the weight on our backs. Just how much are we carrying for others? Are we too quick to go the extra mile and shoulder too much for love, friendship, or service? The wise camel who appears to know his limits advises us to know our own, and to resist carrying more than we can bear.

From docile and sweet to stubborn and vengeful, the camel reminds us that our own mood swings and grudges can make us, at times, hard to handle. Although there are situations when it makes sense to hold our ground, there are probably more times when it's healthier and more productive to let things go and get on with our journeys.

Similarly, when we are dealing with people who have dug their heels into the sand, can't let go of resentments, or only want to do things their own way, it might be good to take a page from the cameleer's handbook and try a more soothing approach. You probably don't need to go as far as singing to such people (though you never know), but some sort of sweetness might break through their camel-like obstinance.

The discerning camel knows his limits and reminds us to know our own, and to resist carrying more than we can bear.

**Confidence • Independence • Liminality • Poise • Predation
Resilience • Self-containment • Wildness**

CAT

THE CAT IN NATURE AND CULTURE

If you live with a cat, you might have wondered how such an independent animal ever allowed herself to become domesticated. Scientists recently shed some light on that question when they discovered that cats didn't yield to domestication as much as they initiated it themselves. About ten thousand years ago, a small wildcat started frequenting human farming communities to hunt the rodents attracted by stored grains. The farmers appreciated the reduction of the pest population, and so they were friendly to the cats. Slowly, over time, the cats warily accepted the invitation to enter our homes, sleep on our beds, and meow when they wanted massages or food.

Ten thousand years later, the wildcat and the domesticated cat do not differ much in their genetic makeup, other than in those genes related to coat colors and patterns. This suggests that the kitty on the couch isn't truly different from the wildcat from whom she evolved. Like her ancestor, the domestic cat still hunts (often returning from her predatory adventures with unwanted trophy "gifts"), wanders (sometimes far from home and for several days), and engages in noisy nocturnal turf wars (occasionally returning with battle scars). No wonder the cat came to represent wildness and the untamed spirit. And her agility, extraordinary sense of balance, uncanny resilience, and ability

After roughly ten thousand years of domestication, cats do not differ much from the wildcats—such as the one pictured here—from which they evolved.

to (usually) land on her feet when she falls gave rise to the folk belief that cats have nine lives.

If a cat is in the mood, she will grace us with her warm, purring presence, granting us the coveted "chosen" feeling that felines are so adept at eliciting. But if she does not find us sufficiently intriguing, the cat will brush us off by turning her back to us to search for something more interesting. When it comes to who's in charge, it is the cat, not us, who determines the rules of engagement. This is why the cat came to symbolize independence, confidence, and self-containment.

As a result of their cool detachment, predatory natures, lightning-quick reflexes, and nocturnal habits, cats have long been associated with the supernatural. Cats were often seen as spirit guides, shape-shifters, and the familiars of witches. Their glowing eyes, superior senses, whiskers that help them find their way through the dark, and ability to move silently on padded feet inspired the idea that cats can travel between the physical and spiritual worlds.

INSIGHTS FROM THE CAT

As every cat owner knows, nobody "owns" a cat. Virtually unchanged from her wildcat ancestors, the cat remained true to herself despite moving in with us. This is what we admire, and even love, about the cat: her ability to hold onto her wild nature while still enjoying the comforts of domesticity and human companionship. In this way, the cat teaches us to know our own natures and retain at least some measure of freedom and independence in our lives. The cat's resounding message is, "Enjoy the relationship, but don't forget who you are!"

Having held onto her wild side, the cat urges us to do the same and give our primal impulses a little more room to play. She whispers, "Go

Cats can connect us to the wild, instinctual parts of ourselves that yearn for the thrill of the hunt, whether stalking an opportunity, opponent, or object of desire.

Even though we may not have the cat's famous "nine-life" resiliency, we can nonetheless let ourselves be inspired by feline fearlessness when it comes to exploring the world.

ahead and pounce on that prattling colleague with a quick and incisive rejoinder." But feline sensibility also tells us that there are times when it's best to retract our claws, climb into a lap, and seductively purr. The mercurial cat therefore shows us that there are many ways to get what we want. She epitomizes flexibility and knows how to land on her feet—both literally and metaphorically. We would do well to learn from her example.

When a cat chases a string, it's likely that she knows the string isn't a mouse—she's simply enjoying the pretense of the moment. Endlessly curious, the cat invites us to investigate the world, even if it means taking chances. The old proverb about curiosity killing the cat is intended to warn humans that too much inquisitiveness can get *us* into trouble. As for the cat, consider the entire proverb: "Curiosity killed the cat, *but satisfaction brought it back*." Having satisfied her curiosity, the cat lives on. The cat, therefore, is a reminder to indulge our curiosities and trust that we will most likely survive our explorations, despite the occasional wear and tear. After all, from the feline perspective, we have lives to spare.

The cat also calls our attention to the discomfort we often feel around creatures who defy categorization, who seem to exist betwixt and between states of being. In his book *The Shaman's Nephew: A Life in the Far North*, Inuit artist Simon Tookoome tells a story about the first time he met a cat—an animal not typically kept as a pet in the Arctic. He was struck by the cat's glowing eyes, sharp claws, and unnervingly poised demeanor. Tookoome was convinced the cat was a "spirit creature" and refused to sleep under the same roof with her. He wasn't the only one ever spooked by a cat. The liminal cat—who has always lived between nature and culture—has a history of sometimes making humans feel uneasy. Perhaps it's because the cat connects us to that dusky place in ourselves where we are neither tame nor wild.

Adaptability • Cleverness • Creativity • Resourcefulness
Survival • Transformation • Trickster

COYOTE

THE COYOTE IN NATURE AND CULTURE

In his book *Coyote America*, author Dan Flores reports that the lore of peoples from the American Southwest claims that the only thing smarter than a coyote is God. If survival as a species has anything to do with intelligence, then coyotes just might possess the incredible smarts these peoples attribute to them.

Despite repeated attempts (historical and ongoing) to exterminate coyotes, they are one of the most widespread and successful predators in North America. Their range stretches from Alaska to Mexico, and their habitats include nearly every kind of environment, from deserts and forests to inner cities and suburban graveyards. When humans moved into and altered their territory, coyotes didn't merely survive the environmental changes—they made the most of them. For example, when we decimated the wolf population, coyotes moved into their canid cousin's ecological niche and began hunting the larger animals who had been previously preyed upon by wolves. As they moved east, coyotes made the ultimate adaptation and bred not only with the shrinking wolf population, but with domestic dogs as well. Recent DNA analysis of Eastern coyotes reveals that their genome is mostly coyote, but also includes Gray wolf, Eastern wolf, and dog. Coyotes always find a way

Both the biological coyote and his mythological counterpart, "Coyote," bend and break rules, cross boundaries, and keep us guessing.

to persevere, which is why they became emblems of wily survival in an ever-changing world.

But it's not just the biological coyote who has proven to be success-ful—the mythological "Coyote" is equally enduring and ubiquitous.

The Coyote character—who appears in the stories of many Native American tribes—is North America's oldest surviving trickster, possibly going back as far as ten thousand years. He has been celebrated and cursed for doing everything, from co-creating the world and bringing fire to the People to causing illness and destroying the buffalo. In his trickster role, Coyote epitomizes the energy of ambiguity and contradiction. He is tireless, hard-working, imaginative, clever, flexible, helpful, and open to new adventures. But he's also lazy, foolish, rigid, deceitful, selfish, and a master of self-sabotage. Coyote shares the best—and worst—of humankind's traits. He often fails, but somehow survives his failures, only to turn around and make the same mistakes again.

In the widespread Coyote story known as "The Bungling Host," Coyote drops in uninvited on an animal friend who has his own unique way of hunting. In the version of the Okanagan people of the Pacific Northwest, retold by Lewis Hyde in *Trickster Makes the World*, Coyote calls on Kingfisher and askes for something to eat. Kingfisher, who uses a specific spearfishing technique, obliges and catches some fish that he serves to Coyote for lunch. Coyote thanks him and returns the favor by inviting him to lunch at his place. The next day, when Kingfisher arrives, Coyote sets out to spearfish using Kingfisher's method, but when he dives into the river—as his friend did—he breaks his neck and dies. Kingfisher goes down to the river, catches some fish, brings Coyote back to life, and lectures him about imitating others. Kingfisher then explains that spearfishing is "his way," and Coyote needs to find his own way. Completely disregarding Kingfisher's advice, Coyote takes Kingfisher's catch back to his lodge and tells his wife and children that he caught the fish. The story ends with Coyote telling his wife that Kingfisher insisted he never fish that way again because it reflects the magnitude of Coyote's power, which is simply too frightening.

At first it might seem like this story suggests that Coyote, lacking his own way of doing things, is distinctly disadvantaged. Even though

When coyote pack members reunite after hunting, they often howl together, creating a high-pitched chorus that reverberates through the night.

he told a lie to make certain his wife never asked him to spearfish again, he's nonetheless left without a way to catch his next meal. And yet, not having his own methods forces Coyote to find other ways to be creative—through his own wit, or more likely through theft or trickery. Adaptability is a recurring theme in Coyote stories, as well as in the natural history of the biological coyote.

The real coyote is a widespread, opportunistic omnivore who can survive on almost any kind of food (from prey animals and carrion to berries and trash). He also has an impressive repertoire of hunting techniques. For small prey, the coyote hunts on his own, conserving energy by slowly stalking and then pouncing. When larger prey is available, he hunts in teams, baiting and chasing the prey until exhaustion wears it out. Sometimes the coyote plays dead, usually in an attempt to lure a vulture, crow, or other carrion-eater within striking distance. Other times the coyote might rise up and "dance" on his hind legs, chase his tail, and behave in other foolish and mesmerizing ways to hold the attention of prey. Meanwhile, his hunting partner sneaks up from behind.

In *Trickster Makes the World*, Hyde relates that coyotes once regularly hunted in packs, similar to wolves. He humorously suggests that eighteenth-century wolves might have chastised coyotes—as Kingfisher does in "The Bungling Host"—by reminding them that pack hunting was the wolf's "way." But two hundred years later, the wolf, trapped in his one way of doing things, is endangered, whereas the ever-adaptable coyote has continuously adjusted to changes in his environment, resulting in a growing population and expanding territory.

Many of the coyote's behaviors reflect his adaptability, including the way he vocalizes. The coyote can vary his voice—by changing its tone and pitch—such that he can sound like an entire pack. In order to deceive both prey and predator, he can even throw his voice by barking into a badger hole. Another unusual coyote behavior is hunting collaboratively with badgers. The unusual relationship is beneficial for both species, as each animal brings a skill the other lacks. The swift coyote can chase the prey above ground and the badger—the champion digger of the animal kingdom—can dive into burrows and dig out the prey if it tries to escape underground. Clearly, the coyote—like his mythological alter-ego—surprises us by breaking rules, tricking others, and adapting to almost everything. Like his trickster persona, he is a consummate survivor.

INSIGHTS FROM THE COYOTE

Seemingly always up for new experiences, the biological coyote shows us that there are many ways to survive in this world if we approach life with creative flexibility. If rabbit's not on the menu, try the berries. If solving a problem isn't working with one approach, try another. Pounce on an opportunity. Distract an opponent by dancing around a situation until you have the advantage. Team up with a friend. Bark into a badger hole. When we feel that we've run out of options, we can find inspiration in the voice of the song dog, who seems to always find a way forward. In contrast with the plaintive sound of a wolf's howl, the coyote's laughter-like barks and yips offer a triumphant message of survival: "I'm still here!"

As for the mythological Coyote, on the surface, he shows us how to behave by misbehaving—he's lazy, vain, and greedy; and he lies, cheats, and steals—but dive deeper and the message he delivers is more complex. He breaks the rules while showing us why we shouldn't break them. He gets away with murder, but he gets what's coming to him. He makes us laugh, yet also makes us cringe. He questions everything, including the questioning of everything. All the while, Coyote is teaching us to accept ambivalence and ambiguity as inherent aspects of life and our own natures. By showing us who we really are—simultaneously bestial and divine, ridiculous and brilliant, cruel and kind—Coyote helps us to stop deceiving ourselves and to stop taking ourselves too seriously. He invites us to question our own ego-driven rigidity and smug self-satisfaction. In doing so, he rescues us from our own worst enemy—ourselves.

The ever-adaptable coyote does whatever it takes to survive, from changing his diet and hunting strategies to breeding with other canids when his own kind are hard to find.

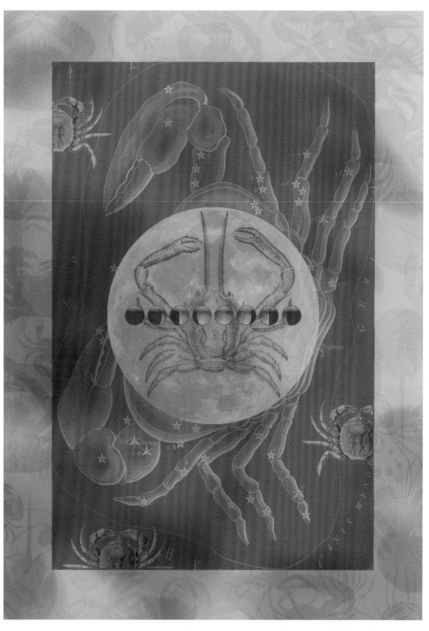

Changeability • Cycles • Rebirth • Regeneration
Renewal • Self-protection

CRAB

THE CRAB IN NATURE AND CULTURE

Wading in the shadows at low tide, we nervously sidestep the scuttling crab who crosses our path. Though the shy arthropod is simply going about his business, his spider-like anatomy and clawed appendages instinctively rattle us, invoking fears of being grasped. Such trepidation gave rise to myths of monstrous crab-like creatures latching onto ships and swimmers, dragging them to the depths of the sea.

Despite knowing that crabs do not pose any threat to us, we can't blame ourselves for feeling a little uneasy around them. The crab is, after all, an unnervingly odd animal. Armored in a protective exoskeleton (carapace), with five pairs of legs (two with claws), and large eyes at the end of long stalks, they are otherworldly creatures. Even the way crabs move—in a frenetic dance of sideways, forwards, and backwards steps—is unusual. The crab's quirky locomotion became a metaphor for something contorted and difficult to understand, as embodied in the adjective "crabbed."

Unquestionably, the crab's claws are his signature anatomical feature, which he uses for grasping, crushing, hunting, fighting, and picking up food. The crab's habit of pinching and grabbing—as well as the tenacious hold of his claws—led to the insults "crab" and "crabby" to describe an unpleasant, unyielding person.

The crab's hard exterior shell invites us to consider the ways we try to hide our own vulnerabilities.

Because of the crab's round shape and tidal habitat, folktales and astrology connect him to the moon. One story from the Philippines, "The Sun and the Moon," relates that the Sun and the Moon were married, and the Moon gave birth to Tambanokano, a giant crab who lives in a pit at the bottom of the sea. When Tambanokano is tucked into his pit, the sea is deep and the tide is high. When he emerges, the sea rushes into the pit, pulling water away from the shore and creating low tide. Tambanokano sometimes becomes so angry with his mother, the Moon, that he tries to swallow her. But the people on the Earth rush out to scare away Tambanokano, so the Moon is saved and returns to her place in the night sky.

Tales such as Tambanokano's reflect the way the moon, tides, and crab are all connected to the cyclical nature of life. The moon waxes and wanes, the tides rise and fall, and the crab advances and retreats. Because the crab must periodically molt, or shed his carapace, as he grows, he is also linked to the ideas of rebirth and renewal. Even more dramatically, when the crab loses an entire limb during territory fights or while warding off a predator, he can regrow it—which further connects him to the idea of regeneration.

INSIGHTS FROM THE CRAB

A creature of both the shore and sea, the crab is both grounded in physicality (shore) and floating in emotional fluidity (sea). He is intimately entwined with the shifting tides—sometimes engulfed by water, other times stranded in the sand—and therefore reminds us of life's ebb and flow. When the water is high, we are either happily immersed or overwhelmingly swamped. When the water recedes, we might feel relieved to be released from the ever-changing currents of emotion or abandoned by the passions that connect us to the world. In this way, the crab offers us an image of life as a circular dance of waxing and waning, ebbing and flowing.

When we resist the ever-changing cycles of our lives, it's often because we are holding on too tightly. Like a grasping crab, we sometimes stubbornly latch onto ideas, perspectives, or other people, refusing to let go, even when holding on is also holding us back or, at the very least, making us crabby to be around. However, when the clingy crab realizes that it's time to let go, he cuts his losses—literally. The crab will lose a limb—or two or three—if that's necessary to avoid capture. Of course, the crab has the amazing ability to regenerate limbs, but doing so takes time and involves risk. Nonetheless, his willingness to lose a limb to save his life can serve as a reminder to throw in the towel and move on when continuing to fight could result in bigger losses.

As the crab zigzags, advancing and retreating, we wonder where he is going or if he is going anywhere at all. The crab's contorted locomotion certainly doesn't get him anywhere quickly. His "crabbed" approach calls to mind those times when it might serve us better to get right to the point. Conversely, approaching things abstrusely does

The crab's tendency to pinch, grab, and tenaciously hold onto objects with his claws inspired the insult "crabby" to describe an unpleasant, unyielding, difficult person.

have its advantages, such as keeping your final destination or objective concealed until you are ready to reveal it.

The crab's coat of armor and claws are impressive defenses, but what lies underneath is a soft vulnerability, reminding us that, when assessing others, we should look beyond surface appearances and consider what might be concealed within. The crab's carapace also calls our attention to the idea of defensive self-protection. Do we create shells to mask our vulnerabilities and prevent others from getting close to us? Do these shells keep us from letting our guard down? Or do they serve as healthy protective boundaries?

The crab's cyclical molting of his carapace can help us remember that we, too, are continually growing and periodically need to cast off our old shells—our personas—to make room for newly developing aspects of self.

Armored in a protective carapace, with five pairs of legs (two with claws), and large eyes at the end of long stalks, crabs are undeniably otherworldly creatures.

**Balance • Cycles • Fidelity • Flexibility • Gracefulness
Journey • Longevity**

CRANE

THE CRANE IN NATURE AND CULTURE

Fossil records tell us that the tallest flying bird—the crane—is also one of the most ancient of all birds, and that most crane species have not changed significantly in ten million years. But long before paleontologists learned this about cranes, Buddhists saw them as symbols of timelessness, as cranes can stand steadily, on only one leg, in running water. As the stream of time rushes past, the crane does not move. She is the calm, still point—unwavering and eternal.

Cranes not only keep their balance while standing on one leg in moving water, but can even sleep through the night in this precarious position. Further, they are considered to be one of the best dancers in the avian world. They are known for their elaborate courtship displays in which they leap, kick, and bow with outstretched wings; arc and lengthen their necks; and lift their heads toward the sky—all in an effort to win over a prospective mate. With moves like this, it's no surprise that cranes became a symbol of balance and gracefulness, and inspired yoga and Tai Chi positions, as well as folk dances.

After such passionate courtship, it seems fitting that most cranes mate for life. Once bonded, they nest, brood, and raise their young together. They also "sing," or vocalize together to strengthen their bond. This is why cranes became symbols of fidelity and marital happiness.

With her fancy footwork and outstretched wings, the crane shows us that sometimes there's no other way to tell the world how we feel except to dance it out.

Outside of breeding season, cranes are gregarious birds, often congregating in large, clamoring flocks. The constant chorus of their calls, combined with the thunder of their wings, has been poetically described as the "voice of the Earth." Perhaps the sound of the cranes' reverberating calls is what prompted the North American Ojibwe Indians to describe them as the "keepers of language," and to name their Crane Clan *Baswenaazhi*, which translates as "echo maker."

Cranes are omnivorous, eating a wide variety of foods ranging from acorns, berries, and leaves to insects, amphibians, and mammals. Because these adaptable birds live a long time—up to forty years in the wild and eighty in captivity—they are a symbol of longevity and luck in China and Japan. The famous Japanese legend about folding one thousand origami cranes to manifest a wish arose from the belief that cranes were messengers capable of carrying prayers to the gods. Cranes seemed especially capable of such a task, as they are able to fly so high (up to twenty thousand feet) that they seemed to touch the edge of heaven. The reason one must fold one thousand cranes (and not fewer or more) is found in the Japanese folk saying that emphasizes—and exaggerates—their longevity: "cranes live one thousand years; turtles live ten thousand years."

Every year, certain species of cranes migrate thousands of miles to nest and feed. When the season is over, they return home, coming full circle. Their annual migration links the crane to the call to the journey and the cyclical nature of life. As the Oglala Sioux medicine man Black Elk said: "Everything the Power of the World does is done in a circle. . . . Even the seasons form a great circle in their changing, and always come back again to where they were."

INSIGHTS FROM THE CRANE

The crane has been on Earth in a relatively unchanged form for millions of years. Her kind has seen species come and go, civilizations rise and fall. The solitary crane standing still in the marsh evokes the feeling of a sentry watching over the passage of time, simultaneously calling to mind the eternal and the transient. Her message to us is to keep things in perspective, let things flow, and not to get stuck in the mud.

An animal's habitat often contributes to the ideas we have about her, and this is especially true for the crane who lives in—and at the boundaries between—three elements: earth, water, and air. Earth is linked to materiality and the body, water to emotion, and air to the intellect. When the crane is in marshes, shorelines, and other wetlands, she is in between earth (materiality and the body) and water (emotion). When the crane is flying just above the water, she is in between water (emotion) and air (intellect). The crane's life in these different realms and her movement through them invites us to think about how we

The crane's annual migration, like those of other birds, connects us to the cyclical nature of life.

balance our physical, emotional, and intellectual needs. Do we tend to prioritize just one realm? Do we spend too much time in the watery world of emotions, forgetting to come up for air? Are we prone to losing ourselves in flights of fancy, even when it's time to come back down to earth?

The crane's flexibility isn't limited to her ability to move through various elemental realms—it also extends to her diet. Like other omnivores, the crane is capable of constantly shifting her focus, sometimes zooming in, when hunting and watching prey; other times taking a wide-angle view, when surveying the landscape for edible plant material. The crane can make a meal of a frog or berries, reminding us that creative adaptability can keep us alive . . . or at least well fed.

To stand on one leg for long periods of time, the crane regularly repositions her body to find her center of gravity. In our own lives—whether we are trying to stay physically balanced when doing yoga or striving to stay emotionally centered when life's challenges overwhelm us—we can take inspiration from the crane. The perpetually poised crane shows us that staying centered requires periodically checking in with ourselves—physically, emotionally, and mentally—and calmly shifting our positions as necessary.

Given her ability to balance, it's not surprising that the crane is considered the most accomplished dancer in the animal kingdom. The crane dances to court mates, strengthen bonds, claim territory, teach her young, and to simply express emotions or have fun. Hopping from foot to foot, twirling, strutting, leaping, and curtseying, the exuberant crane reminds us that sometimes there's no other way to tell the world how we feel except to dance it out.

Finally, the crane's migration can be looked at as analogous to the hero's journey; an archetypal pattern involving three basic stages. First, there is a call to adventure. For the crane, this is her drive to migrate. Next, a journey that tests the hero's resolve. This is the crane's arduous flight to her wintering grounds, which is risky, but will reward her with a chance to find a mate, nest and birth chicks, or simply feed and survive. The final stage is the hero's triumphant return home, having been changed by the journey. Although we cannot know how the crane who returns in the spring has changed, it seems fair to assume that she has experienced many things and so could not possibly be the same bird she was when she departed in the fall. In this way, cranes beckon us to answer the age-old call to embark on a journey of transformation.

**Chaos • Creativity • Deception • Endurance • Ferocity • Fertility
Fury • Instinct • Primal • Survival**

CROCODILE

THE CROCODILE IN NATURE AND CULTURE

Consider the classic crocodilian ambush strategy: the stealthy reptile lies at the water's edge, as still as a log, until her prey is within reach. She then lunges, closes her massive jaws on an unsuspecting victim, and drags him underwater until he drowns. Now she has a choice to make: swallow her catch whole or, if it's too large to eat all at once, store the carcass under rocks or logs for future consumption. If it's egret nesting season, the crocodile might use a different technique, one involving a lure. She will float with twigs on her head, looking like a rock covered in debris. When a foraging egret in search of nesting material tries to collect a twig, the crocodile rises from the water and snaps her jaws shut on the unsuspecting bird. No wonder the crocodile symbolizes deceit and ferocity.

When we see crocodiles (as well as their close relatives, the alligator, caiman, and gharial), it feels as if we are looking at a creature from the time of the dinosaurs. These timeless reptiles haven't changed all that much in more than two hundred million years because they developed such successful adaptations to their habitat and ecological niche. They have thick, leathery skin covered in scales that protect them from predators and even the impact of bullets. With nostrils, eyes, and ears conveniently positioned along the top of their heads, they can breathe,

Female crocodiles are protective, nurturing mothers who keep their young safe by carrying them in their jaws—a place no predator would dare approach.

smell, see, and hear even when most of their bodies are hidden underwater. When completely submerged, their ears and nostrils close, and their eyes are protected by an additional set of clear lids that function as goggles.

The crocodile also has incredible endurance. Combining swimming with riding the ocean currents, the saltwater crocodile has been known to travel hundreds of miles by sea. Another crocodilian adaptation is her ability to reduce her heart rate to just a few beats per minute, which enables her to stay underwater for nearly an hour. And her throat is equipped with bony flaps that allow her to eat underwater without drowning. When it is time to rest, the crocodile sleeps with one eye open and one half of her brain active and on the lookout. As for her bite, the saltwater crocodile has a bite force of 3,700 pounds per square inch (psi). To put this in context, lion and tiger bites generate only about 1,000 psi.

The crocodile's incredible adaptations link her to the concept of survival, while her primitive appearance connects her to the idea of the primordial world. The Yolngu, an Aboriginal Australian people, believe that they are descended from an ancestral crocodile, and so consider crocodiles sacred. Because of the crocodile's strength and endurance, she also is seen as an emblem of power and leadership, which gave rise to the Tabwa people of Central Africa referring to crocodiles as the "lions of the water."

Despite the force of her bite and her dragon-like fierceness, the female crocodile is actually a very gentle, nurturing mother. She protects her young for up to two years, even tenderly carrying them inside her powerful jaws when necessary to keep them safe. And when her hatchlings call, she comes running. Because of her maternal behavior, as well as her association with water, the female crocodile is a mother archetype and linked to fertility and creativity.

Like other animals who live both in water and on land, crocodiles are creatures of the borderlands. They lurk at the muddy edges between rational, conscious materiality (earth), and irrational, unconscious formlessness (water). As such, crocodiles are associated with the primal energy of both creativity and destruction as reflected in the myth of Sobek, the Egyptian crocodile god who emerged from the dark water of chaos and created order in the universe.

INSIGHTS FROM THE CROCODILE

Crocodiles are survivors with an arsenal of defenses, including thick, nearly bullet-proof skin, which might have inspired the idiom "thick-skinned" to describe someone with emotional resiliency. So, if one is rejected by a prospective date, the crocodilian response would be to toughen up, let it go, and look for another prospect. Having a thick skin helps us to not take things personally and, with time, leads us to feeling more comfortable in our own skins.

The crocodile is a master of deceit and concealment, staying hidden until her targets are within reach.

When it comes to hunting strategies, the crocodile has several pointers. First, she sleeps with one eye open, always ready to respond to prey (as well as the occasional predator), demonstrating that sometimes all it takes to succeed is simply staying alert. Second, she is a self-preserving opportunist; she doesn't waste a lot of energy searching for prey. Instead, she lurks in the water, waiting for prey to come to her. (Interestingly, the word "lurk" is used as a noun in Australia—a place known for crocodiles—where it describes an underhanded scheme or a lazy way of performing a task.) Finally, whether lurking or waiting out a time of scarcity (she can go a year or longer without food), the crocodile seems to have all the time in the world. She waits to strike until her targets are within reach, modeling the benefits of patience and precision timing. If we take her lead, we wait to ask for the promotion until after we deliver a stellar presentation.

The crocodile is also a master of concealment, misdirection, and camouflage. When silently floating in the water, still as a log, she can fool biologists and birds alike. Covering herself with twigs, she combines the use of bait and pretense like the best con artists. Regardless of whether we like to admit it, deceit and concealment—as when one bluffs in a poker game—can be an effective strategy. So beware the con artist with a toothy leer.

We conclude with the image of a crocodile dragging us under the water and tearing us apart as a metaphor for those times when we feel incredibly overwhelmed by the circumstances of our lives. During such challenges, we can feel dragged into the watery depths of emotional turmoil and shaken up so fiercely that the water muddies and we can't make our way out. But the crocodile—who uses her jaws to rip through a five-hundred-pound wildebeest *and* to tenderly carry her hatchlings—reminds us that life is an endless dance between creative and destructive forces. Out of chaos comes order, and with time the silt settles, the waters clear, and we find our way back onto the riverbank, where we can bask in the clarifying light of a new day.

Elusiveness • Gentleness • Regeneration • Seduction
Sensitivity • Spirituality

DEER

THE DEER IN NATURE AND CULTURE

The Sami reindeer herders of northern Europe speak a language that includes more than one thousand terms describing reindeer, reflecting how important these animals are to their culture. Their high regard for reindeer is further revealed in one of their creation stories, in which the world was created from the body of a reindeer. The Gwich'in people of Alaska and Canada have a similarly poignant origin myth that tells how people and caribou were once one being, who shared the same heart. When this being was separated into caribou and human, caribous kept a little piece of the human heart, and humans kept a little piece of the caribou heart, which keeps them forever connected.

Deer are one of the most widespread—and most hunted—mammals on the planet, with forty-three different species distributed over every continent except Australia and Antarctica. Since prehistoric times, deer have provided humans with meat and milk for sustenance, hides and sinews for clothing and tents, and bones and antlers for tools, weapons, and other implements. But just as important as what deer have provided on a material level is the impression they made on the human imagination. From their appearance in paleolithic cave paintings to the part they play in mythologies and traditions around the world, deer have always loomed large in the human psyche.

Many cultures regard the deer as a spirit guide who leads humans into spiritual realms or brings them closer to divinity.

There is something about the presence of a deer that triggers a primal response and beckons us to follow. When a deer crosses our path, it is nearly impossible not to stop, meet his gaze, and feel a sense of awe. And when he vanishes in the blink of an eye, we feel the urge to take a few steps toward where he had been, and perhaps a few more if we see tracks. Before long, the deer has led us out of our familiar world and back into the proverbial forest—a place of wild uncertainty that symbolizes the unconscious and initiation. The pursuit of a deer into the woods then becomes a metaphor for a journey that can change us on a profound level.

Because the deer can lead us in this manner, and because he can seemingly appear and disappear at will, many cultures have perceived him as a spirit guide with supernatural powers. For example, in Siberian shamanic traditions, shamans sometimes enter trances and spiritually ride to the sky on a reindeer. Similarly, the Huichol—an indigenous people of northwest Mexico—regard the deer as a guide who acts as an intermediary between gods and shamans. And within the Celtic tradition, deer are seen as messengers from the otherworld and are linked to the fairy realm.

Once deer became seen as messengers from other realms, the pursuit of deer took on the symbolism of a spiritual quest. In Sir Thomas Malory's *Le Morte d'Arthur*, King Arthur repeatedly attempts to capture a white stag, but the deer always eludes him. In these stories, as well as many others, the stag represents the pursuit of spiritual wisdom. In Christian legend, the stag sometimes represents the call to God. There are stories about two hunters—Hubertus and Eustace—who each encountered a magnificent stag with a crucifix standing between his antlers. The experience caused both men to give up their worldly possessions and become pious Christians.

In Buddhism, deer symbolize the Buddha's teachings, as well as the act of receiving them. The dharma wheel, or *dharmachakra*—one of the oldest and most important symbols of Buddhism—is typically depicted between two deer. When Buddha gave his first sermon in a deer park, where he set Buddhist doctrine (dharma) in motion, deer peacefully gathered around and listened to him speak.

Generally timid and gentle, the doe is a symbol of the feminine. Native American lore of the Eastern Woodlands and Central Plains tribes tells of a seductive Deer Woman who shifts between human and deer forms, sometimes helping women to conceive, other times luring adulterous men to their death. In contrast, the stag, who symbolizes masculinity, strength, and pride, has been depicted as the king of the forest. As a result of his cyclical antler shedding and regrowth, the stag is also linked to regeneration and vitality. Because the doe and stag each epitomize aspects of their respective sexuality, and because both sexes entice hunters to follow them, deer are linked to seduction.

INSIGHTS FROM THE DEER

The deer who appears to be peacefully grazing is nonetheless vigilant. At the snap of a twig under a hunter's boot, he is in motion. The hunter may turn toward the sound of the hoofbeats, but more often than not, he doesn't even catch a glimpse of the deer. Perpetually aware of his environment, the deer demonstrates how staying attuned to our surroundings enables us to sense threats and slip away. On the other hand, the hypervigilant "prey animal" mindset can cause us to see everything as a potential threat, which fosters fear and limits the scope of experience.

When the seductive deer tempts us to follow him into the forest, he is symbolically asking us to leave behind the safety of the familiar. He leads us into the wilderness, but makes no promises about what we might find—or lose—as a result of following him. In this way, the deer connects us to the call to adventure, prodding us to embark on a transformative experience in which we might learn something new about the world and ourselves. Do we follow him into the unknown—the metaphorical forest—or turn back home to the comfort of the familiar?

Every year, in late spring, male deer (except for reindeer, the only deer species in which females also have antlers) sprout a new set of antlers that will grow all summer. At first, the antlers are covered in a hairlike material known as velvet. By the end of summer, the deer sheds his velvet covering, revealing a fully grown set of hard, polished antlers just in time for the autumn breeding season. Come winter, the antlers start to weaken; by early spring, they are shed, and the cycle begins again. The seasonal shedding and regrowth of antlers reminds us that we, too, go through cycles of endings, in which we have to let things go, and new beginnings filled with the promise of what's to come.

Because the doe and stag each epitomize aspects of their respective sexuality, and because both sexes entice hunters to follow them, deer sometimes symbolize the idea of seduction.

Community • Devotion • Friendship • Guardianship
Loyalty • Protection

DOG

THE DOG IN NATURE AND CULTURE

In the story "The Cat that Walked by Himself," by Rudyard Kipling, the first man awakens to find a dog in his cave, and he asks the first woman, "What is Wild Dog doing here?" The woman responds, "His name is not Wild Dog anymore, but First Friend because he will be our friend for always and always and always." As it turns out, Kipling was right when he located the origin of the human-dog relationship in the cave-dwelling days of the Paleolithic. Current research suggests that dogs and people have been friends for about thirty thousand years. Scientists speculate that around this time, wolves loitering around human hunting camps were welcomed for their ability to provide warnings about approaching predators or to assist with tracking prey. Over time, these wolves evolved into dogs, and the dog became not only our *first* friend, but also our *best* friend.

Since those early days around the fire, dogs have been our friends in countless ways. They have helped us hunt, herd livestock, guard our homes and businesses, catch criminals, sniff out explosives, search out and rescue missing people, assist and guide those with disabilities, and more. They do all this because of their ability to enter into relationships with other species. Like wolves, dogs are social animals, and given the choice they will join a pack, which for most dogs is their human family.

Dogs inspire us to come to our senses—to return to an embodied, primal consciousness based on visceral experience, not abstraction.

This ability to develop strong social bonds with humans and other animals have made dogs nearly universal emblems of friendship. And the qualities that they demonstrate, such as constancy and devotion, have made them one of the most enduring symbols of loyalty and faithfulness.

With keen senses that enable them to detect things that humans cannot, dogs also represent instinct, intuition, and perception. They can see in the dark, hear nearly twice as many frequencies as we can, and have a sense of smell somewhere between ten thousand to one hundred thousand times more sensitive than our own. As a result of their superior senses, they can track, hunt, and navigate much better than we can. Dogs, therefore, have been seen as mediators between the human and animal worlds, between culture and nature, consciousness and the unconscious, life and death. This is why dogs were once considered guardians of the gateway between worlds.

INSIGHTS FROM THE DOG

Dogs can guide us back to things that matter. For example, when we walk our dogs, they sniff, listen to, taste, and touch nearly everything they encounter. In this way, dogs inspire us to come to our senses—to return to an embodied, primal consciousness based on visceral experience, not abstraction. As humans, it is often too easy to get lost in our thoughts and not be present in the moment. If we were to follow the example of our dogs, we might actually stop and smell the proverbial roses. This is because, for dogs, life is all about the here-and-now. Sadly, for most of us, it often takes a serious crisis before we remember to live for today. Thankfully, dogs can help us to realize that worry won't spare us any sorrow tomorrow, but it will sap today of its joy.

One of the big differences between dogs and humans is that dogs don't repress their emotions or deny what's going on around them. Yes, they use coping strategies (such as not making provocative eye contact with unfamiliar dogs), but they don't lie to themselves about the dog not being there. When we find a chewed-up sneaker and look at our

The dog's capacity for love and devotion has made him an enduring symbol of loyalty.

Humans and dogs have been "best friends" for roughly thirty thousand years.

dogs, they don't pretend that they didn't do it. Instead, they lower their heads in what looks like contrition. Although our human world admittedly requires us to practice a certain amount of repression and pretense in order to cope with the constraints of culture, most of us would benefit from expressing more of the uncomplicated honesty that dogs do.

Dogs also help us to remember how much family and friends matter. Work deadlines, financial pressures, and other distractions often prevent us from realizing—let alone expressing—how much other people mean to us. Like dogs, we are pack animals, and most of us really do need a pack to be happy. But, unlike dogs, we often fail to treat other people in ways that reflect our affection for them. Dogs are always happy to see their family members and friends, and they always express this happiness. This is one of the reasons we like dogs so much—they never let us forget how much they love us. Imagine how much kinder the world would be if we could do the same for one another.

Current research suggests that what distinguishes dogs from other animals is their extraordinary capacity to form affectionate relationships with members of other species.

DAUPHIN RHINOCÉROS. ⅍. , DAUPHIN ALBIGÈNE. ⅍.

Altruism • Breath • Communication • Community • Perception
Playfulness • Sociability

DOLPHIN

THE DOLPHIN IN NATURE AND CULTURE

In her book *Beautiful Minds*, biologist Maddalena Bearzi tells a story about a group of dolphins who were feeding on a school of sardines when one of the dolphins abruptly swam away at a fast clip, with the other dolphins following behind. Bearzi found the behavior odd, so she turned her boat around and pursued them. Three miles offshore, the dolphins suddenly stopped swimming and formed a circle. As Bearzi approached, she discovered a young woman floating, near death, at the center of the circle. Bearzi took her to the hospital, where she learned that the young woman had intended to commit suicide by drowning. Her identification, along with a suicide letter, were sealed in a plastic bag tied around her neck. The doctors told Bearzi that if she hadn't found the young woman when she did, she would have drowned.

The dolphin has long been associated with altruism and selflessness because there are so many similar stories—both historical and contemporary—of dolphins helping humans without any incentive or promise of reward. Some dolphins guide ships to safety or assist fishermen; others save people from drowning or sharks. The dolphin's good deeds led Celtic peoples to see her as a guardian of the water. Within the Christian tradition, the dolphin's altruistic behavior was perceived as virtuous, so she was often depicted as delivering the souls of the faithful

By virtue of their altruistic behavior, dolphins encourage us to summon the better angels of our own natures.

to Christ. Similarly, in ancient Greece, the dolphin was believed to carry the souls of the dead to the Islands of the Blessed.

Perhaps we shouldn't be surprised by the dolphin's altruistic behavior, given their social intelligence. To start with, dolphins appear to be self-aware. They recognize their own reflections in a mirror, and use signature calls that function as names for themselves and other dolphins. Dolphins seem to have a theory of mind: the ability to sense what others are thinking. They also demonstrate empathy: the ability to understand and share the feelings of another. They cooperate with one another, engage in teamwork, teach their young, and form communities (pods). Dolphins, therefore, have come to represent community, sociability, and communication.

The dolphin is also linked to the idea of perception through echolocation—the ability to see the world through sound. Dolphins send out sound waves and, when they hit an object, bounce back, giving the dolphin information about the location, shape, and size of the object. Scientists have known for a long time that echolocation enables dolphins to navigate, communicate, hunt, and protect themselves from predators. More recent research suggests that dolphins can also see *into* objects, similar to the way ultrasound imaging technology can see into our bodies. Based on experiments and anecdotal evidence, it appears that dolphins also can use their echolocation to detect pregnancy or cancerous tumors. It's even possible that dolphins might be able to use echolocation to read emotions, which might explain why the dolphins described at the start of this chapter reacted the way they did to the woman who attempted to take her own life.

Like whales and porpoises, dolphins breathe air. When they dive underwater, muscles close their blowholes, which stay closed until they surface and exhale. Some species of dolphins can stay submerged for more than ten minutes. Their ability to control their breathing and spend long periods of time underwater links dolphins to breath, which symbolizes life and spirit.

Dolphins are the pranksters of the sea and will sometimes play catch with sea turtles, tossing the unwitting turtles around like balls.

Like humans, dolphins thrive when they are sociable and suffer when they are on their own, and so can serve as a reminder to tend to our own "pods" and make time to connect with our friends.

And when it comes to spirit, dolphins have plenty—especially when it comes to their *joie de vivre*. They mate for more than just reproductive purposes, suggesting they enjoy sex for pleasure. Perennially playful, they create and manipulate underwater bubble rings, engage in "keep away" games with strands of seaweed, and perform acrobatic leaps and spins, to name only a few examples. They also are known as the pranksters of the sea and will sometimes drag birds under water without any apparent intent to eat them, tease fish by offering them bait and then snatching it away, and play catch with sea turtles—seemingly just for the fun of it.

INSIGHTS FROM THE DOLPHIN

By virtue of their altruistic behavior, dolphins encourage us to summon the better angels of our own natures. When a colleague is drowning in work, we could lend a hand and lift them up. If a friend is "lost at sea," we could take the time to guide them back to shore. Conversely, if we are always playing the savior role and trying to rescue everyone in our lives, we need to take a different page from the dolphin handbook and lighten up. The puckish dolphin frolicking in the waves seems to know better than most of us that all work and no play makes Flipper a dull girl. She would advise us to seize the day.

Like humans, dolphins thrive when they are sociable and suffer when they are on their own, and so can serve as a reminder to tend to our own "pods" and make time to connect with our friends. Furthermore, when it comes to play, dolphins engage not only with one another, but even with other species—such as humans, dogs, and whales—thereby building bridges across the species divide. Do we have the same ability to reach across the divides of differences?

Because echolocation enables dolphins to see with sound, they can use it to find their way in murky and deep waters, where there is neither clarity nor light. In such conditions, they may not be able to see with their eyes, but nonetheless find their way in the dark. This idea offers us a metaphor for being able to see our way through emotional or spiritual darkness, suggesting that, during challenging times, we might need to find new ways to navigate. Perhaps instead of relying on the rationality associated with "seeing clearly," we need to take a more intuitive approach in which we are guided by other ways of understanding ourselves and our situations.

When we watch dolphins swimming and playing underwater, it can be hard to remember that they are air-breathing mammals like us. But they are, and they need to regularly surface to breathe. We, too, need to periodically come up from the depths of our immersion in life's chaos to catch our breath and regain our perspective. Then, like the dolphin, we can dive back in—renewed and once again ready to ride the waves.

Agility • Illusion • Light • Magic • Perception • Transformation

DRAGONFLY

THE DRAGONFLY IN NATURE AND CULTURE

Snake Doctor. Devil's Darning Needle. Water Witch. Meadow Hawk. Jacky Breezer. These are just a few of the many whimsical nicknames for insects in the order *Odonata*—the creatures commonly known as dragonflies and damselflies. Around the world there are hundreds of different regional folk names for these fascinating creatures who have existed on Earth for roughly three hundred million years.

Of all insects, few have engaged our imaginations the way dragonflies do. Nicknames such as Sewing Needle and Devil's Darning Needle reveal old folk beliefs that these insects would sew shut the mouths of impudent children, nagging women, irreverent men, and liars of any kind. Names such as Snake Doctor and Adder's Servant reflect the related belief that these "flying sewing needles" stitched the wounds of injured snakes.

But not all perceptions of the dragonfly were negative. Some saw these harmless insects in a kinder light, and nicknamed them accordingly. In Serbia, for example, the dragonfly is known as *vviilliin kkon-jjiicc*, meaning "fairy's little horse," and in parts of England and Europe they are sometimes called Water Butterflies.

Given their unusual appearance, it's not surprising that dragonflies inspired so many stories. They are large, visually striking insects with

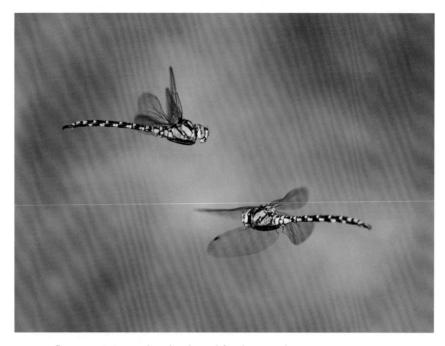

Dragonflies sometimes dart back and forth over the same area—a motion similar to sewing—which inspired folk beliefs about dragonflies stitching shut the mouths of those who misbehave.

distinctively elongated bodies and two sets of delicate cellophane-like wings. Their large eyes cover most of their heads and enable them to see nearly 360 degrees at all times. When dragonflies alight near us, they compel our attention, and often seem to respond to our gaze by turning their heads as if regarding us with similar interest. With their shimmering iridescent coloring that changes depending on the angle of viewing or illumination, they seem to be part insect and part trick of the light, which is why they are associated with illusion and magic.

Often described as a "master aerialist," the dragonfly can dive, dart, hover, fly sideways or backwards, and can exceed speeds of thirty-five miles per hour. She sometimes flies back and forth over the same area, in a motion similar to stitching, which no doubt led to the folk beliefs related to dragonflies sewing humans and other animals. Because of the dragonfly's remarkable flying abilities, as well as her time spent flying high in the sky near the heavens, she is linked to the idea of communication with the gods, as well as to the souls of the departed.

Despite her ethereal beauty, the dragonfly is a formidable fighter and has been known to attack and take down hummingbirds. In Japan, where the dragonfly is known as *katsumushi*, or "victory insect," she appears as a recurring motif on the armor, helmets, and weapons of Samurai warriors. Similarly, for the Plains Indians of the United States and Canada, dragonflies are symbols of invincibility, and their image was painted on war shirts, shields, tipis, and other objects as protection against injury.

The lifecycle of dragonflies has made them a nearly universal symbol of change and transformation. They begin their lives as wingless aquatic

Dragonflies don't acquire their spectacular forms easily—they go through a metamorphosis in which they can molt as many as fourteen times in order to transition into a beautiful winged insect.

The dragonfly's iridescent coloring changes based on the angle from which she is viewed, which offers us the idea that shifting our point of view can sometimes have a profound impact on how we see the world.

creatures (called nymphs) with gills and then go through a metamorphosis in which they can molt as many as fourteen times in order to complete their transition to dragonflies. During their time as nymphs, which can last as long as a few years, they develop wings and go through other changes. As they prepare for their final molt, known as the "emergence," they begin to breathe air. Finally, they leave the water, break out of their last larval shell, and emerge as flying insects. The adult stage lasts only weeks or months, during which time they mate.

INSIGHTS FROM THE DRAGONFLY

When you look at a dragonfly, do you see a Devil's Darning Needle (sinister intent) or a Water Butterfly (ethereal beauty)? Our response to the dragonfly asks us to consider how we perceive other creatures and what we project onto them.

The dragonfly's iridescent coloring—created by the play of refracted light on their bodies—shows us the value of looking at life from multiple angles. Shifting our point of view can have a profound impact on our experience. When we are stuck in a static pattern or worldview, we can miss all kinds of insights and opportunities. We approach life more creatively when we are willing to stay open to the full spectrum of possibilities.

Like many animals who undergo a transmutation, the dragonfly starts life in one form and goes through radically different stages before reaching maturity. A dragonfly experiences this life cycle only once. But for us, change is an ongoing, ever-unfolding process. We may never reach a single "emergence" as the dragonfly does, but if we are open to change, we are more likely to discover and realize our potential.

The dragonfly spends most of her short life "becoming." By the time she reaches her adult winged form, she has only a few weeks to live. But during those weeks, the dragonfly mates, sometimes in flight. A male and female join their long bodies together, forming a flying heart-shaped "mating wheel," providing us with a symbolic and poignant reminder to make the most of our time.

Domesticity • Emotional Comfort • Foolishness
Humor • Sensitivity

DUCK

THE DUCK IN NATURE AND CULTURE

"If it looks like a duck, swims like a duck, and quacks like a duck, then it probably *is* a duck." This amusing maxim expresses the idea that something can usually be identified by its habitual characteristics, and that there's no need to come up with other theories. We could express the same idea by using almost any animal, but the person who coined the phrase most likely chose the duck because there is something about this unassuming waterfowl that is charmingly straightforward. We might even say that a duck is what a duck does, given that the English word "duck" was first a verb meaning "to dip or plunge suddenly," and only later came to be a noun used for certain birds.

Besides ducking, the other characteristics that make a duck a duck start with a compact, streamlined body covered in tightly layered waterproof feathers that serve as insulation. Her legs are positioned toward the back of her body and her feet are wide and webbed. Both of these characteristics make her perfectly suited to paddling and diving. However, those swimmer legs and feet also give her an awkward gait—known as a waddle—which many people find amusing.

One of the duckiest things about ducks is their signature vocalization: the quack. Ducks quack to communicate with each other, and when ducks are gathered together, one quacking duck quickly turns

The duck's seemingly easy-going nature, along with the way water rolls off her back, can remind us not to take things too seriously.

into a chaotic chorus. It's not a stretch to see why this image has been applied to a group of friends (usually women) gossiping. There are other metaphoric links as well. When someone is speaking foolishly, he might be said to be "quacking" about something. The words quack (and quackery) refer to someone who holds outlandish opinions, or who is a huckster, fraud, or charlatan. This association holds true in other languages, not just English. For example, the French idiom for a false or misleading report, rumor, or story is simply the French word for duck, *canard*. All of these associations are based on the prejudice that ducks are not to be taken seriously, which is why ducks are associated with silliness and have inspired so many children's stories and toys.

However, there's more to ducks than their waddling and quacking. Like other waterfowl, they are masters of multiple realms: earth, water, and sky. They are equally at home foraging on the shore, paddling in a lake, flying at high speeds (up to sixty mph), great distances (eight-hundred-mile migrations), and at high altitudes (up to four thousand feet). Perhaps it is the ease with which ducks move through different realms that led various cultures to give them a role in creation stories. For example, the Native North American people of the Crow Nation say that Coyote sent Duck to the bottom of the ocean to bring up mud from which Coyote created the land. Similarly, a Hungarian tale recounts how Magyar, the Sun God, turned himself into a golden duck and flew from heaven to the bottom of the sea, where he retrieved the seeds that grew into animals and people.

As a migratory bird who tends to return to the same territory every year, and often mates for life, the duck is linked to the concepts of stability, fidelity, and domesticity. A common behavior of duck mothers is to line up their ducklings close behind them, a behavior from which we get the expression "to get one's ducks in a row" (meaning to get everything organized and prepared). One of the reasons that ducklings follow their mother so faithfully is their imprinting, a brain process by which they become socially bonded to the first moving object they see after hatching—essentially saying, "That's mom." Because of this imprinting behavior and their overall nurturing ways, ducks have come to symbolize emotional comfort and sensitivity.

INSIGHTS FROM THE DUCK

One day, a single duck quacked, other ducks readily joined in, and whomever was listening thought, "Reminds me of gossiping, and the way rumors are started." Although we are unlikely to ever know what ducks are quacking about, these vocal birds can nonetheless prompt us to ask ourselves, "What are *we* quacking about? Is it worthy of repetition or sharing on social media? Or are we mindlessly repeating just another . . . *canard?*"

When a duck surfaces after a dive underwater, she emerges looking dry because duck feathers are coated in a water-resistant oil. This

In order to watch over her ducklings and keep them from getting into trouble, a mother duck often lines them up close behind her, a behavior that provides us with a metaphor for keeping things organized.

water repellency gave birth to the idiom "rolled off him like water off a duck's back," meaning that something has no significant impact on someone. So, when things are getting to us, we need to take a page from the duck's playbook and figure out how to shake off whatever it is that dampens our spirits.

Ducks imprint during their first few days of life when they are in what is called their "sensitive period." During this time, whomever they first see makes a strong and lasting impression on their brains. Though not as dramatic as a duck's sensitive period, we nonetheless go through times (such as in a rebound phase after a break-up) when we are more receptive to whomever might show up in our lives. At such moments, we might ask ourselves: "Am I imprinting on the *right* person, or the person who just happened to come along at the right time?"

The guileless duck suggests that we not take ourselves too seriously. She paddles through life with an unpretentious ease. Hers is an unsinkable buoyancy—literally and metaphorically—and the lesson she offers is this: sometimes all it takes to stay afloat in life is floating. Whether peacefully bobbing along, diving to feed with her butt in the air, waddling and quacking with her pals, or getting her ducklings in a row—a duck *is* what a duck *does*, and much of what she does is sweet, silly, and just ducky. As the poet Frank W. Harvey wrote, "From troubles of the world I turn to ducks: Beautiful comical things."

Divinity • Dominion• Duality • Light • Power
Predation • Sovereignty • Vision

EAGLE

THE EAGLE IN NATURE AND CULTURE

At the top of a giant tree—a tree so big that the entire universe is held in place by its branches and roots—sits an eagle whose flapping wings create the winds of the world. At the bottom of this tree, coiled in the tree's roots, is a serpent, who is in perpetual conflict with the eagle. These two creatures, positioned at opposite ends of the world, keep all of creation in balance. This mythological motif—an eagle and snake in a sacred tree that holds the world together—is widespread and found in stories and art from many cultures around the world. It also tells us a great deal about how humankind sees eagles.

When an eagle soars at great heights or descends toward prey at speeds of more than one hundred miles an hour, he looks regal and powerful, which is why he so often has been seen as the king of birds and as an emblem of sovereignty and dominion. Eagles fly at altitudes of more than ten thousand feet and build their nests, known as aeries, high above the ground at the tops of trees or on cliffs. Riding thermal updrafts on a clear day, an eagle can appear to graze the edges of the sun and yet remain unscathed. For this reason, ancient peoples believed that the eagle could look directly into the sun's fiery light without damaging his eyes. The Gaelic name for the eagle, *iolaire sùil na grèine*, or "eagle of the sunlit eye," is a poetic reflection of the eagle's solar connection.

The word raptor comes from the Latin word *rapere*, meaning to seize or take by force, and that is exactly what the eagle does.

Pliny the Elder, the first-century Roman naturalist, attributed another supernatural power to the eagle: an immunity to lightning. The eagle's ability to survive contact with lightning also is found in stories about the Greek sky god, Zeus, who assumed the form of an eagle in order to hurl thunderbolts. An image of Zeus in his eagle form, clutching a thunderbolt in his talons, even appeared on ancient Greek coins. A similar association between lightning and eagles is found throughout Native American mythology. For many tribes, the eagle was the inspiration for the mythical thunderbird, a powerful force of nature who possessed the ability to control thunder with his wings and lightning with his eyes.

Perceived as capable of withstanding the heat and light of the sun, controlling wind and weather, and flying so high they seemed to touch heaven itself, eagles were further presumed to have a special relationship with the gods. For example, in numerous Native American cultures, the eagle delivers messages to the creator, and eagle feathers are considered sacred, and even used in prayer and ceremony.

The eagle's association with light and divinity link him to wisdom, as does his incredible visual acuity. Scientists have determined that, with the eagle's 340-degree range of vision and their ability to zoom in on details from afar (an eagle can see a rabbit from a distance of three miles), seeing the world through an eagle's eyes would be like simultaneously looking through powerful telephoto and wide-angle lenses. But even before scientists learned about the eagle's eyesight, people understood that there was something special about his vision. In 65 BCE, the Roman poet Horace wrote, "Those who are eagle-eyed in spotting others' faults are blind to their own." Today, we still use the expression "eagle-eyed" to describe someone who is sharp-sighted and quick to notice even the smallest details.

The eagle is a raptor, and like all raptors, his superior eyesight, powerful talons, sharp hooked beak, and strength make him a skilled and

When we try to look at our lives from the eagle's perspective—the quintessential birds' eye view—we are more likely to see the larger patterns that shape our lives.

formidable hunter. The word raptor comes from the Latin word *rapere*, meaning to seize or take by force, and that is exactly what the eagle does. He seizes and kills his prey with his talons—which have a crushing force of roughly four hundred pounds per square inch—and then tears it apart using his beak. Although eagles excel at hunting, they also are opportunistic and not above stealing prey or scavenging.

Even the predatory and noble eagle has a soft side. Eagles will sometimes perform aerial stunts by themselves or play with found objects, seemingly just for fun. Most eagles mate for life, and they begin their courtship with acrobatic nuptial flights in which a male and female will soar, dive, spiral, and spin, often while linking talons. Both parents build their nest together (often returning to the same spot year after year) and take turns incubating their eggs and tending to their young.

INSIGHTS FROM THE EAGLE

With his incredibly acute vision and elevated vantage point, the eagle can see the big picture as well as the smallest detail. In this way, he teaches us to not only pay attention to the little things, but to also expand our vision so that we can be attentive to the full scope of the larger patterns that shape our lives. Once we have a clearer view of our landscape, it is easier to avoid obstacles, set priorities, and pursue that which calls to us.

The sight of an eagle soaring in the sky evokes the urge to fly high, to accomplish and achieve, and to rule our domain. So when life offers us those rare moments of updraft, we should go for it, fly like an eagle, and enjoy the sense of freedom from our usual limitations. Conversely, the single eagle silhouetted against the sky reminds us it can be lonely at the top. For those of us on the ground, it can be difficult to relate to someone who is soaring above us. So even when we are flying high, we need to remember to come back down to earth on occasion.

Furthermore, flying too high has its risks, as the Greek myth of Icarus tells us. Despite warnings not to fly close to the sun with his

The way the eagle appears to soar so close to the sun inspired the ancient belief that he could look directly into the sun's fiery light without damaging his eyes.

wings made of feathers and wax, Icarus was so taken with his newly acquired power of flight that he flew too high. The heat of the sun melted his wings and he fell into the sea and drowned, leaving us with a warning about arrogantly disregarding our own limitations.

Perhaps the most profound insight that the eagle offers us concerns life's dualities. The eagle is symbolically connected to both the golden light of the sun and the dark shadows of thunder clouds. He mates for life and tenderly cares for his young, but will readily tear the flesh from still-living prey. Because he soars so high, we imagine him as a divine courier, carrying our prayers to heaven. But we can also find him hunched over a carcass on the side of the road, gulping down bits of carrion, eyes flashing with the ferocity of nature. The eagle, therefore, shows us that life manifests as both light and dark, clarity and chaos, and, ultimately, matter and spirit. Just as the eagle and serpent in the cosmic tree hold creation together, it is only when we learn to accept life's counterpoints that we can begin to feel unified and whole.

**Community • Compassion • Largeness • Loyalty • Memory
Resoluteness • Stability • Strength • Wisdom**

ELEPHANT

THE ELEPHANT IN NATURE AND CULTURE

Everything about the elephant is big—his body, mind, heart, lifespan, and the impression he leaves on the human imagination. As the largest land animal on Earth—African males can weigh as much as seven tons and stand twelve feet at the shoulder—the elephant is synonymous with immensity. This is why when something is huge, gigantic, or enormous, it often is compared to an elephant, or simply described as elephantine. Because of his massive size, the elephant symbolizes strength and stability to such an extent that in Hindu mythology the Earth is said to be supported on the backs of elephants (who in turn are supported by a giant turtle).

With such heft, these colossal creatures are capable of making an impact on the world merely by moving through it. When elephants travel through a landscape looking for food and water, they often trample underbrush and topple trees. A grazing elephant can consume hundreds of pounds of vegetation in a day. In their quest to eat and drink, they create pathways and access points not only for themselves, but for other animals, as well. This probably explains why, in the Hindu pantheon, the elephant-headed god of wisdom, Ganesha, also is known as the remover of obstacles and is associated with good luck and success.

As Ganesha's association with wisdom suggests, the elephant isn't all brawn and no brains. Considered one of the smartest of animals,

The elephant reminds us to let ourselves be "big," to find ways to remove obstacles in our path, and to make space for ourselves in the world.

elephants engage in tool use, such as using branches to reach for food and to scratch their backs. They cooperate to solve problems, and they remember one another—as well as humans and other animals—over many decades. (Elephants can live as long as seventy years in the wild.) It even appears that older female elephants—the matriarchs, or "grandmothers"—may pass on their acquired knowledge to younger herd members in a kind of cultural transmission. The elephant's impressive memory made him a symbol of the wisdom that comes from remembering what one learns during a long life. It also led to the saying "An elephant never forgets." For the Ashanti people of central Ghana, elephants were regarded as reincarnated human leaders who returned in elephant form to govern the other forest creatures with their wisdom and discernment.

Although the elephant's physicality and intelligence are impressive, perhaps the most notable qualities associated with elephants are their kindness and compassion. Elephants care for and protect one another by providing assistance to ill, injured, and trapped herd members. They have been seen holding one another up when injured, rescuing one another's calves from river currents and predators, and removing darts and spears from one another's bodies. Elephants also are known for acting altruistically toward other species. These sensitive creatures even perform ritual-like behaviors around newly dead elephants and over the bones of long-dead elephants, suggesting that they have an understanding of death and may experience grief.

All these qualities of heart have made elephants paragons of kindness who compassionately care for one another. The benevolence of these gentle giants is celebrated in the Buddhist tradition, where they are associated with spiritual strength—and even with the Buddha himself. One story relates that when the Buddha's mother was pregnant with the Buddha, she dreamt of a white elephant who offered her a white lotus.

Elephants are empathetic animals who are known to console one another during times of stress or when injured.

In elephant society, older matriarchs keep their herds away from danger, help them find food and water, and teach the younger females how to care for their offspring—all of which results in a healthier herd.

INSIGHTS FROM THE ELEPHANT

The elephant can teach us to let ourselves be "big," to find ways to remove obstacles in our path, and to make space for ourselves in the world. There is no point—or pleasure—to be found in playing small, for when we do we are less likely to reach our full potential or find ways to make the world a better place. In nature, as the elephant moves through a landscape, it is his enormousness that creates pathways and opportunities for everyone in the forest. Conversely, the elephant can sometimes cause destruction by over-grazing and trampling, so he also can serve as a reminder to pay attention to the ways in which we impact the world while seeking to achieve our goals.

Conspicuous in his bulk, the elephant inspired the expression "the elephant in the room," a metaphor for an uncomfortable issue that one tries to ignore despite its obvious presence. As such, the elephant calls our attention to just how powerful human denial can be when it comes to not wanting to recognize a troubling issue, even when it looms large right in front of us.

Always willing to lend a helping trunk, the elephant teaches us about the importance of caring for all members of a community. Even unrelated elephants attend to the ill and injured, sometimes at their own expense, which makes the herd stronger than if each elephant were out only for himself.

The long-lived elephant "who never forgets" can teach us to value the wisdom gained over a long life, while shining a light on how we regard our elders. In elephant society, the older matriarchs keep their herds away from danger, help them find sources of food and water, and teach the younger elephants how to care for their offspring. Apparently, a matriarch's knowledge is so valuable that herds with older matriarchs tend to fare better than those without. Like elephants, when we honor the experiences, knowledge, and skills of our elders, we increase our chances of a better future. Just as a herd with older matriarchs tends to be healthier and stronger, a human society that learns from previous generations is more likely to thrive.

Adaptability • Cunning • Ingenuity • Seduction
Swiftness • Trickster

FOX

THE FOX IN NATURE AND CULTURE

Often described as the most catlike canid, foxes defy categorization. Like cats, they usually hunt alone, stalking and pouncing on their prey in a distinctly feline style. They have catlike eyes with vertical pupils that help them see in low light, and long, sensitive whiskers that help them sense their environment. One species, the gray fox, even has semi-retractable claws that enable him to climb trees. Despite these cat-like characteristics, foxes—with their small social groups, dominance hierarchies, and omnivorous diets—are true canids. In fact, they are the most widespread wild canid in the world.

Incredibly adaptable, foxes live in fields, forests, mountains, deserts, tundras, and even urban environments. Foxes are also exceedingly opportunistic, and will eat just about anything. Consummate omnivores—rivaled perhaps only by people in their lack of culinary discernment—foxes are happy to dine on small mammals, birds (including chickens), reptiles, frogs, eggs, insects, fish, mollusks, fruits, vegetables, fungi, carrion, and even trash.

Since ancient times, people have admired the fox for his ingenuity, cunning, and slyness—and for good reason. Foxes are superb strategists. When pursued during a fox hunt, this crafty canid might confuse the hounds by doubling back on his own trail or by leaping onto a sheep to

The fox plays a role as trickster in hundreds of folktales, from Aesop's Fables in ancient Greece to the Reynard stories of medieval Europe and the Brer Fox lore of the American South.

camouflage himself with her scent. When the fox is the one doing the hunting, he relies on a different set of tricks to catch a meal. He might squeak like a rat or bleat like a lamb in order to deceive and lure his next meal. On other occasions, the fox might play dead to draw a crow or other carrion-eater close enough to pounce on. The wily fox will even use his tail to brush away his own footprints in the snow. Foxes demonstrate impressive forethought, which is where we get the expressions sly as a fox, cunning as a fox, and to outfox.

Often seen at threshold times, such as dawn and dusk, and in areas where wild and cultivated environments transition into one another, foxes came to be associated with boundary zones and the magic of liminal spaces. These elusive canids seem adept at finding secret pathways, slipping through impenetrable thickets of brush, and taking underground shortcuts that appear to defy spatial logic. This may be why

Incan, Siberian, Celtic, and other cultures consider the fox to be a spirit guide who helps shamans and spiritual seekers find their way.

According to many old stories, the fox can shape-shift into other animals and even humans. For example, in Japanese *Kitsune*, or fox stories, the fox is portrayed as a smart, supernatural being who can transform himself into a human. One of the ways the *kitsune* accomplishes his metamorphoses is by pulling his fiery flame-shaped tail between his legs and rubbing it with his forepaws until it sparks and flashes into a human. Once in his human form, the fox sometimes plays the role of an irresistible seducer or seductress.

The fox also plays a role as trickster in hundreds of folktales, from Aesop's Fables in ancient Greece to the Reynard stories of medieval Europe and the Brer Fox lore of the American South. In many of these stories, the fox outsmarts less wily animals and people, and has great fun at the expense of others. But not all folktale foxes are naughty;

The wily fox will sometimes use his tail to brush away his own footprints in the snow, demonstrating impressive forethought.

The fox reminds us to pay attention to the person trying to go unnoticed, as he might be trying to outfox us.

some are kind-hearted and help people who are struggling with a difficult situation, or use their cleverness to help lost people find their way.

INSIGHTS FROM THE FOX

Symbolically, the fox has a lot to offer our imagination. Because he lives in the boundaries between cat and dog, day and night, wild and cultivated, trickster and helper, the fox reminds us that things are rarely black and white, but instead more nuanced. Within the dusky ambiguity of ever-shifting shades of possibility, the creative and flexible fox warns us that we cannot be too stuck in our ways if we want to be successful, happy, or even just able to move forward. Instead, we need to stay open to possibilities and be willing to adapt, improvise, and make the best use of whatever resources and strategies we have.

The fox often epitomizes subtlety. A master at blending into his surroundings, he is usually the spy—not the one being spied upon. He knows how to watch, wait, and . . . learn. The elusive fox doesn't show his hand until it suits him to do so. Vulpine strategy therefore cautions us to pay attention to the person trying to go unnoticed, as he might be quietly planning to outfox us. Conversely, we might try the same approach ourselves: hang back, observe, and bide our time until we have the edge.

When subtlety doesn't work, the nimble-minded fox has other tricks up his sleeve. He might try a technique sometimes referred to as "charming," in which he engages in mesmerizing antics—such as leaping, rolling, and chasing his own tail—to lure curious prey closer so that they can get a better look at what he's doing. When they are close enough, the fox pounces on them, proving that there's no fool like an outfoxed fool. The fox's behavioral camouflage should feel familiar to us primates, for we do it all the time, such as when trying to surprise or deceive in both work and play. For better or worse, the fox shows us that pretense has its place.

In his trickster form, the fox warns us that we are all capable of beguilement. Even those of us not easily seduced have our weaknesses. Whether captivated by a romantic prospect (or simply a compelling scheme), we sometimes fall prey to a slick presentation. So, when a rakishly handsome fox—or a too-good-to-be-true opportunity—appears in our lives and invites us to throw caution to the wind and run with him, beware of his foxy charms!

Evolution • Fertility • Reaction • Rebirth • Renewal
Sensitivity • Song • Transformation • Water

FROG

THE FROG IN NATURE AND CULTURE

Have you ever wondered why the prince is so often turned into a frog in fairy tales? After all, storytellers could have just as easily chosen skunks, slugs, or other unappealing animals to play this role. Folklorists and psychoanalysts have speculated about all sorts of reasons for casting the frog as a prince, but the most convincing is his easily observable metamorphosis. The frog begins life as a limbless aquatic tadpole with gills and then turns into a limbed air-breather who can live on land as well as in water. His striking transfiguration makes him a powerful symbol of transformation. Factor in the frog's somewhat humanlike appearance—a smooth, hairless body; legs longer than his arms; and front-facing eyes—and it's no wonder that he was cast as the slimy suitor who shape-shifts between frog and human.

As creatures who spend much of their lives in ponds, lakes, swamps, and other freshwater habitats, frogs are associated with water. They often croak before it rains and emerge from the water afterward, creating the appearance that frogs and rain—as well as the fertility rain brings—are interconnected. Within the North American Zuni tradition, frog fetishes are buried near water sources to keep them flowing. Similarly, the Aymara people of the Andean highlands place frog effigies on mountaintops to manifest rain. Frogs are also linked to fertility through their enthusiastic

The frog is the personification of readiness—always tuned in and ready to leap into action.

mass mating, which can turn an otherwise placid swamp into a clamorous croaking orgy. During mating season, males will mount almost anything—other males, fish, or even a human hand. It's easy to imagine the ancient Romans, after observing the frog's passion, then deciding that these ardent amphibians must be sacred to Venus, the goddess of love, who was sometimes depicted in the company of frogs.

Because they are prolific and produce large quantities of eggs, frogs also are linked to fecundity and abundance. This may be why the ancient Egyptian pantheon included a frog-headed goddess of childbirth known as Heqet, and Egyptian women wore frog amulets to bring them success in childbirth. Perhaps the most notable evidence that the ancient Egyptians saw the frog as a symbol of abundance is their hieroglyph for the number 100,000, which is a tadpole.

Producing lots of eggs is one way to ensure survival for these soft-skinned, vulnerable creatures. Another way is to evolve a heightened sensitivity that enables a quick response to threats and opportunities—which is exactly what frogs have. To start with, the position of their eyes (on top of their heads) provides them with a nearly 360-degree view of their surroundings, helping them to detect even the slightest movements of both predator and prey. Then there's their signature move—the jump. As one of the most impressive athletes in the animal kingdom, frogs can leap up to 150 times their own body length. They also have lightning-quick responses that enable them to jump out of harm's way—often before they are even seen by a predator. But their tongues are even faster. A frog can cast out his sticky whip-like tongue, capture an insect, and reel it back into his mouth about five times faster than a human can blink.

As a result of their trigger-fast reflexes, frogs often leap away from us before we can see them—but that doesn't stop us from hearing their peeps, croaks, ribbits, and other calls, especially in the early spring. Their annual vernal concerts have made them a symbol of rebirth and renewal, as has their hibernation cycle. During winter, frogs hibernate, some underwater and others on land in leaf litter or sheltered nooks. Even when the temperature drops below freezing, frogs never freeze completely because they have a high concentration of glucose in their vital organs. Parts of their bodies will freeze, and they might even stop breathing. But when the weather warms up, frogs' bodies thaw, and, as if by magic, they start moving again, seemingly returning to life.

INSIGHTS FROM THE FROG

Because the frog developed earlier than other classes of life (such as mammals), and because he progresses from a lower (tadpole) to higher (frog) state of being, the frog is often seen as a symbol of evolution. Within a few months, the frog goes through the kind of transformation—a complete change in body plan—usually associated with long spans of evolutionary time. As humans, our fundamental anatomy

Folklorists and psychoanalysts have speculated about all sorts of reasons for casting the frog as a shape-shifting prince in fairy tales, but the most convincing is his easily observable metamorphosis from tadpole to frog.

won't change during our lives, but our personalities and perspectives will. There will be times when, in order to make room for an ever-developing self, we will need to completely change our forms, metaphorically speaking. Such transformation isn't easy. The frog who turned into a prince had to do the hard work of convincing a fussy princess to allow him at her table—and in her bed—before he could transform. Often, before we can change, we have to convince others—or even ourselves—that there's more to us than meets the eye.

Frogs go for it—with gusto. When it's mating time, they don't hold back. The males sing their "pick me" mating calls until an entire pond seems to vibrate with their fervor. They don't spend time weighing the pros and cons of croaking the way we might deliberate over the efficacy of online dating. Frogs put themselves out there because that's the only way they will have a shot at finding a mate. They show us that sometimes we just have to take a chance in order to bring anything to fruition, especially romance.

Always alert to changes in his surroundings—such as the arrival of a predator, potential prey, or even a mate—the frog is the personification of readiness. He shows us the benefits of being tuned in and ready to leap into action. By being more fully engaged, we are better able to reap benefits and avoid threats. With a heightened sense of what's going on around us, it's easier to be flexible and make adjustments when situations call for it, or detect opportunities and advance toward them.

The frog's sensitivity to his environment is not merely sensory—it's also physiological. Considered an "indicator species," frogs tend to be one of the first organisms to show signs of distress when a habitat turns toxic. In this way, frogs shine a light on the interconnectedness of life and the importance of paying attention to changes that might be forewarnings. When frogs start vanishing, it signals that we need to take action and work toward restoring environmental health. Similarly, when we see signs of negative changes in our personal "ecosystems"—whether in physiological, emotional, social, or other arenas—we should take note and make reparative steps while we still can.

Interestingly, the Japanese word for frog, *kaeru*, also means "to return." Perhaps this dual meaning was inspired by the frog's association with spring, which always returns, or his ability to seemingly return to life after hibernation. Either way, in Japanese, *kaeru* is often paired with the word for money in the idiom, *okane ga kaaeru*, which expresses the idea of money or good fortune returning. As an indicator species, frogs remind us that if we heed warnings, we can potentially turn things around, thereby inviting frogs—and prosperity—to return.

Agility • Balance • Capriciousness • Exuberance
Fertility • Gregariousness • Impertinence
Independence • Stubbornness • Vitality

GOAT

THE GOAT IN NATURE AND CULTURE

Anyone who has spent time with goats will enthusiastically tell you that they are remarkable creatures. They are intelligent, curious, gregarious, and fearless. Like dogs, domesticated goats know their names, wag their tails, follow people around, and even appear to have a sense of humor, based on their behavioral response to human laughter. For example, when goats act out in front of people who laugh at them, the goats are more likely to intensify their flamboyant antics. Goats seem to truly love an audience.

Endlessly playful when young, goat kids will run in circles, jump vertically in the air, leap over other goats, whirl in place, dance, prance, and engage in all kinds of frisky play. As adults, they are gifted climbers with a graceful gait and an upright tail—the picture of physical confidence, sure-footedness, and agility. With their incredible sense of balance, they can climb sheer rock faces like the best mountain climber, and some goats even climb trees to reach desirable fruit. Everything about goats—from their exuberance and spirited sense of play to the lustful behavior of bucks in rut—suggests that they have a uniquely goatish *joie de vivre*, which gave rise to their symbolic association with such qualities as vitality, fertility, and sensuality.

Goats will butt heads to play, spar, or to establish dominance, the latter of which inspired the expression, "butting heads," to describe a heated confrontation.

Archeological sites provide evidence that people started raising goats as long as ten thousand years ago. Although we don't know what prompted the first people to domesticate the goat, it might have been their good nature. That said, goats aren't always easy to keep. There is a popular Finnish proverb that warns, "If you're short of trouble, take a goat." Goats can be temperamental and impertinent. Often unwilling to take direction, goats sometimes refuse to move, or even eat, if a goat keeper tries to coerce them to do something they don't want to do. Furthermore, these intelligent animals are easily bored and are very good at finding clever ways to escape their enclosures. Unlike sheep, who tend to stay close to their herd, goats are happy to wander off on their own to find new grazing grounds. Goats, therefore, came to epitomize independence and stubbornness.

Historically, the goat's capricious, rebellious, and lustful behaviors were in direct opposition to a chaste and ordered society, which is why the goat inspired the half-goat, half-human forms of Dionysus and the satyrs in Greek mythology, Bacchus and the fauns in Roman mythology, and Satan in the Judeo-Christian tradition. Making the goat a symbol of socially unacceptable behaviors was itself a form of "scapegoating," the act of blaming the wrong entity (an innocent person or group) for the wrongdoings of others. The origin of the word scapegoat comes from the ancient Jewish tradition of designating a goat as the carrier of the sins of the people and then banishing it to the desert.

With their incredible sense of balance, goats can climb sheer rock faces like the best mountain climbers, and will even climb trees to reach desirable fruit.

INSIGHTS FROM THE GOAT

Much of the time, it seems like goats just want to have a good time—playing, climbing to the top of everything in their path, breaking free of constraints, and exploring the world. What's not to love about their free-spirited nature? While the sheep stay in the pasture and follow the shepherd, the charismatic rebel of the barnyard teaches us to have some fun, forge our own path, and seek new heights. We might even want to break the rules now and then.

While there's nothing like the comforts of home and herd, the goat invites us to expand our horizons. The goat is not satisfied with the same old paddock. He will climb a mountain to graze on a new flavor of shrub or go out on a limb—literally—by climbing a tree to nibble on fruit. The sure-footed, adventurous goat shows us that independence, curiosity, and agility have their rewards. Confidence helps, too. The goat gets his, at least in part, from his innate sense of balance, suggesting that when we are centered and balanced, we have more confidence and competence.

The goat's capricious behavior can connect us to the unpredictable and sometimes chaotic power of creativity. Before we can create something new, we often need to go through a disorderly phase in which we try out different ideas or processes. The impulsive, free-wheeling goat can lead us into the chaos from which creativity and order often arise. And the ever-passionate buck in rut can serve as a not too subtle reminder that our sensual natures need room to play, too.

As the original scapegoat that humanity ritually burdened with its sins, the goat prompts us to think about blame and forgiveness, and whether we take responsibility for our failures and shortcomings or project them onto others.

Charming as goats can be, they also can be stubborn, surprisingly picky, and unapologetically self-motivated, so they can shine a light on these same qualities in ourselves. Even a goat knows that digging in and holding one's ground isn't always worth the consequences; there are times when yielding is the better strategy. Obstinance often keeps us stuck on one opinion or course of action, whereas yielding sometimes sets us free to get on with life, climb the next mountain, and reach for the most luscious fruit at the very top of the tree.

Goats are intelligent, curious, gregarious, fearless, and even appear to have a sense of humor. Their *joie de vivre* is contagious and can inspire the same in ourselves.

Compassion • Cycles • Devotion • Fidelity • Fellowship
Fertility • Journey • Loyalty • Teamwork

GOOSE

THE GOOSE IN NATURE AND CULTURE

The Latin maxim *Omne vivum ex ovo* declares that "All life comes from an egg." This idea is at the heart of numerous origin stories that tell of a celestial egg from which the entire universe hatched. As to which came first—the divine egg-layer or the divine egg—it depends on the myth, but in at least a few stories the cosmic egg-layer is a goose. For example, in Egyptian mythology, a celestial goose—known as The Great Honker—laid the original egg from which all life emerged.

In Aesop's fable "The Goose that Laid the Golden Egg," the eggs aren't origin points for all of creation, but are nonetheless extraordinary. In the tale, a farmer acquires a goose who lays a golden egg each day. Believing the goose to be full of gold, the farmer is overcome by greed and slaughters the bird, expecting to find a belly full of golden eggs. Instead, he finds nothing but the insides of a goose. The story illustrates the risk of overreaching greed, and gave rise to the idiom "killing the goose that lays the golden eggs," referring to the short-sighted destruction of a valuable resource.

Cosmic eggs and golden eggs connect the goose to the concept of fertility, as does her association with goddesses, such as Aphrodite, the Greek goddess of love, and Juno, the Roman goddess of marriage and childbirth. And let's not forget Mother Goose—of nursery rhyme

Geese are monogamous, maintain tight family units, and are fiercely protective of their young and their mates.

fame—who is thought to have her origins in much older Germanic goddesses who were depicted with flocks of geese.

Given that geese are linked to divinity and gold, it's fitting that their natural history offers us an example of living by the golden rule (treating others as you want to be treated). As members of a gaggle (a group of families who form a community), geese look out for one another by taking turns watching for predators while others feed and signaling the group if there is a need to take flight. During migration, if a member of the flock is injured or becomes ill, two other geese leave the migrating flock and follow her to the ground, where they stay with the injured goose until she can fly again, or until she dies.

Another example of the goose's "golden-rule" behavior can be found in the their flying habits. Geese fly in an aerodynamically advantageous "V" formation, in which the geese in front reduce air resistance so those flying behind use less energy. Geese take turns moving from the front of the formation to the back so that they all share the work. As a result of observing these golden-rule behaviors in geese, we associate them with qualities such as loyalty, compassion, and fellowship.

Geese also have strong family values. They are monogamous, maintain tight family units, and are fiercely protective of their young and their mates. Males forage for food and will often feed their mates before themselves. Like ducks, newly hatched geese imprint—or bond with and devotedly follow—the first moving object they see. For these reasons, geese, like ducks, have long been symbols of faithfulness and devotion.

Many birds migrate, but geese in particular have come to symbolize the archetypal journey: their autumn departure represents the call to adventure and their triumphant return in spring heralds a promise fulfilled.

INSIGHTS FROM THE GOOSE

Because the goose had the honor of laying the world egg, from which all creation emanated, she invites us to think about creativity and what we are incubating in our own lives. A goose will often roll a bad egg out of her nest rather than waste precious energy on it, suggesting that we ask ourselves whether what we are trying to hatch is worthy of our effort.

However, once a goose decides that her eggs are healthy, she and her mate become seriously invested in their welfare. Both parents are very caring and protective, reminding us to prioritize whatever it is that we are incubating and nurturing in our own lives. The imprinting behavior of hatchlings—the way they will follow the parental figure everywhere—is an especially poignant image of the responsibility we bear when leading those who are most impressionable. However, the idea of imprinting also cautions us to be alert to relationships driven by timing, vulnerability, or biological drives rather than suitability.

Geese aren't just devoted to their young, but are also devoted to other geese—starting with their mates. The pair-bond that mated

When geese migrate, those flying in front reduce air resistance so those flying behind use less energy. They take turns moving from the front of the formation to the back so that all the geese share the work.

geese forge awakens thoughts about the tender and intimate space that any two devoted creatures can create. Within the larger community of their gaggle, geese are quintessential team players and call our attention to the value of working together. The V-formation of their migratory flights allows geese to travel much farther together than they ever could on their own. Similarly, when we share the burden—by dividing the work, taking turns, and periodically changing our positions or roles—we often get the job done more efficiently and with less effort. Geese are the epitome of the motto "all for one, and one for all." They never leave behind one of their fallen friends, thereby providing an inspiring model of a society based on reciprocal compassion. The world would be a better place if we all behaved more like geese and stood by one another in hard times, even when it requires personal sacrifice.

The goose's cyclical fall migration calls our attention to the ever-turning wheel of time. Heralded by the sound of their calls as they once again take flight, their annual departure reminds us that the nights are lengthening and another year will soon come to an end. This poignant marker of the passage of time provokes us to think about the path of our own lives. The geese's flight—in an arrow-shaped chevron formation—points decidedly forward, and prompts us to ask ourselves, "Where am I headed?" and "What new directions are beckoning?"

**Ferocity • Focus • Freedom • Soaring • Vision
Warfare • Watchfulness • Wildness**

HAWK AND FALCON

THE HAWK AND FALCON IN NATURE AND CULTURE

A falconer raises her gloved hand, calls out toward the treetops, and a hawk with a three-foot wing span lands on her arm. Despite our widespread familiarity with the sport of falconry, the image of a raptor responding to a human's command seems incongruous. After all, birds of prey aren't the sort of animals we think of as tame, and the truth is they aren't. Although humans have been breeding, raising, and hunting with hawks, falcons, and eagles for at least six thousand years, these creatures have never been truly domesticated. This is one of the reasons why these magnificent birds are icons of unconquerable wildness.

All flying birds symbolize freedom, but especially hawks and falcons, because they soar—both literally and metaphorically. To conserve energy, they spread their wings and let air currents—thermals and updrafts—carry them higher and higher without needing to flap their wings. When there isn't an air current, they tuck in their wings and move through the air like an arrow until they catch another thermal or updraft. Because these birds ascend in the sky, they do the same in our imaginations, which is why our spirits *soar* during moments of great happiness and why someone *soars* to new heights when they achieve extraordinary accomplishments.

Although humans have been hunting with hawks and falcons for at least six thousand years, these creatures have never been domesticated, which is why they symbolize unconquerable wildness.

Another arena in which raptors soar above other birds is hunting. The term "raptor" comes from the Latin word *rapere*—to seize or take by force—which describes the way hawks and falcons dive out of the sky to catch prey. Like all raptors, hawks and falcons are skilled aviators with powerful talons and sharp beaks. The Peregrine Falcon is the fastest animal on earth, and has achieved diving speeds of more than two hundred miles per hour. Hawks are lightning fast, too, often descending on prey at speeds of one hundred and twenty miles per hour. Because of their powerful physical attributes and the fierceness with which they capture their prey, both hawks and falcons are linked to the concepts of ferocity, predation, and warfare. The use of the word "hawk" as a term for people who support war comes from this symbolic association.

Raptors have excellent binocular eyesight, and some species—such as the Red-tailed Hawk—can see a mouse from a distance of one hundred feet. It's no wonder that when we pay very close attention to someone, studying their every move, we are said to be watching them like a hawk. These birds are the epitome of focused alertness and the ability to see what is otherwise hidden.

As creatures of high altitudes who appear to fly close to the sun, hawks and falcons have been seen as messengers between heaven and Earth, the companions of gods, and even gods themselves. In the ancient Greek pantheon, hawks were sacred to Apollo, the sun god, and several ancient Egyptian gods were depicted with falcon heads, most notably Horus, the sky god. One of the most fascinating mythical raptors can be found in Australian Aboriginal stories about "fire-hawks." These birds were said to hunt with fire and were credited with bringing fire to humans. As it turns out, hawks and other raptors are indeed drawn to fire. When there is a blaze, hawks and falcons capture prey fleeing from the smoke and flames. Scientists have known this for

When hunting, a raptor focuses intensely, descends like a missile, and in the blink of an eye she is flying away with prey in her talons—a paragon of expediency.

As a result of their perspective and acuity, hawks and falcons can read the terrain, symbolizing the ability to see into a situation and interpret it.

a long time. Recently, however, indigenous folklore converged with science when researchers discovered that certain raptor species go one step further: they pick up live embers, carry them to brush or grass outside the fire zone, and drop them in what appears to be an attempt to start a new fire to flush out even more prey.

INSIGHTS FROM THE HAWK AND FALCON

It's nearly impossible to watch hawks and falcons fly and not feel envious. As they ride the air currents, they are the quintessential symbol of freedom, evoking the idea that the sky's the limit. Raptors also demonstrate that there's more than one way to stay aloft. When in flight, rather than constantly flapping their wings, they make the most of the thermals and updrafts to fly with the least amount of effort, offering us a lesson in working smart by going with the flow.

From their high vantage point, hawks and falcons direct our attention to the value of big-picture thinking. As a result of their perspective
and focus, these birds can read the terrain, symbolizing the ability to
see into a situation and interpret it. Similarly, when we take a bird's
eye view of things, it is often easier to maintain our objectivity, which
enables us to see patterns and make connections. That said, a persis-
tently lofty, above-it-all perspective overshadows the value of humility,
subjective experience, and the practical lessons learned from having
our feet on the ground.

When hunting, a raptor focuses intensely, descends like a missile,
and in the blink of an eye she is flying away with prey in her talons.
She gets right to the point in order to achieve her goal, illustrating the
value of expediency. Her approach prompts us to consider our own
strategies. Do we dive into a situation in order to make things happen
for ourselves, or do we linger too long on a comfortable perch, hoping
things will come to us? On those astonishing occasions when raptors
use fire to flush out prey, they metaphorically suggest that sometimes,
in order to get things done or bring things to light, we need to apply a
little heat. Conversely, despite the raptor's admirable ability to seize the
day, not every occasion calls for laser focusing or dive bombing with
hawkish intensity. There will always be times when it's better to conjure
one's inner dove instead.

Finally, we return to the image of the falconer and her raptor, which
embodies one of the most interesting symbolic aspects of these birds: the
tension between wildness and tameness. When hunting with humans,
raptors are completely free and could easily fly away. Occasionally,
when released, a trained bird will fly off but, more often than not, she
returns to the falconer because she trusts her and associates her with
sustenance. This unpredictable dance of release and return—or freedom and attachment—is at the heart of the falconry relationship, and
ultimately at the heart of all our relationships.

**Community • Cooperation • Creativity • Fertility
Productivity • Reward**

HONEY BEE

THE HONEY BEE IN NATURE AND CULTURE

The honey bee is nature's tiny alchemist. She transforms nectar, pollen, and water into the liquid gold we call honey. Then she stores the honey in tiny, hexagonal containers made of wax. Honey bees are the only animal, other than humans, who combine materials in a way that produces an entirely new substance. Mammals produce milk and silkworms produce silk, but both milk and silk are created within these animals' bodies. Only the honey bee—the culinary artist of the animal kingdom—gathers, combines, and manipulates materials outside of her body in order to manufacture a new substance. Furthermore, the substance honey bees make—honey—isn't merely delicious (it has often been described as the nectar of the gods), but also has medicinal and antimicrobial properties, which make it that much more extraordinary.

Because of their prolific honey production, bees have become symbols of productivity, creativity, wealth, and the sweetness of life. But honey isn't the only thing bees produce. One-third of the human diet comes from insect-pollinated plants, and the humble little honey bee is responsible for pollinating 80 percent of these plants. As a result of their extensive pollination, honey bees also symbolize fertility.

Honey bee hives are paragons of order. There are only three classes of bees, and each class has a specific task or sets of tasks. The hive is

Honey bees are associated with creativity because they are the only animal, other than humans, who combine materials outside of their bodies in a way that produces an entirely new substance.

ruled by a single queen, whose main purpose is to reproduce. The all-male drones have one primary duty: mate with the queen. The workers, who are all female, have two responsibilities. They act as nursemaids, tending to the queen and larvae, and as foragers, leaving the hive to find the pollen, nectar, and water that sustains the hive. This clearly defined hierarchy made bees a perfect metaphor for human civic order, caste systems, and monarchies. For example, in ancient Greece and Egypt, the monarch's emblem was a bee.

The bee's work ethic inspired the idiom "busy as a bee," and there's no doubt that their constant activity bears this out. The bees of a typical hive need to fly about ninety thousand miles—that's roughly three flights around the Earth—to collect just over two pounds of honey. The daily buzzing of their incessant activity and productivity made bees symbols of industry throughout history and across cultures.

Honey bees are dormant in the winter and re-emerge in the spring, and so are associated with the concepts of reincarnation and resurrection. Plato believed that the souls of the virtuous were reincarnated as honey bees, which gave rise to the honey bee as a symbol of the soul, reflected in the expression "to tell the bees," meaning to communicate with the departed.

In order to communicate certain kinds of information to one another, worker bees return from their foraging and perform a series of movements. The direction and duration of these movements, known as the "waggle dance," correspond to the location of nectar, pollen, water sources, and potential locations for new hives. The more excited the forager bee is about what she discovered, the more she waggles in an attempt to convince the other bees to check it out.

INSIGHTS FROM THE BEE

We might not be able to turn sunlight into sweetness like the honey bee, but this remarkable insect can inspire us to think about the process of bringing things together to create something new. As they fly from flower to flower, fertilizing the plants that produce sustenance for so many, these tiny pollinators also prompt us to consider what we are fertilizing or bringing to fruition in our own lives.

When bees find nectar, they don't waste any time making the announcement. They make an immediate "beeline" directly back to the hive to alert the other foragers so as not to risk losing a valuable resource. In doing so, they demonstrate the value of focus and expediency. When the forager bees arrive and do their waggle dances, they vigorously shake their bodies from side to side and buzz their wings. They are creating a "buzz" about a good place to eat, or about the presence of rivals or predators who might cause harm to the hive. Something similar happens when we discover a promising new restaurant and can't stop talking about it, or see a

Plato believed that the souls of the virtuous were reincarnated as honey bees, which gave rise to the honey bee as a symbol of the soul.

film that disappoints us—our "buzzing" creates or negates interest. Bees, therefore, call our attention to the incredible power of word of mouth.

Honey bees teach us about the sweet rewards of making an individual effort *and* working cooperatively. They also remind us of the merits of belonging to a community and contributing to the common good. But the community-oriented hive mind can also have its downside. When a bee stings to warn off an intruder, her stinger releases pheromones—chemical signals that attract, arouse, and provoke other bees to swarm in a massive defensive response. But the act of stinging kills the bee, so many lives are lost. In this way, bees warn us of the risks of the hive mind in our own lives and in society. We risk losing the clarity of our own choices and values when we allow ourselves to be pulled into groupthink—whether we're under the sway of peer pressure or subsuming ourselves to the norms of our culture.

Most of us are afraid of stinging insects, such as bees, and when we run in fear we send out a chemical signal that actually draws the bees to us. But if we remain calm—as most beekeepers do—honey bees will rarely sting. Through the way they react to our fear, bees provide an opportunity for us to learn how to control our emotions and cultivate serenity in our lives. The rewards of such composure yield sweetness, not only in the form of honey for the beekeeper, but also in the form of inner peace—and fewer stings—for the rest of us.

**Desire • Empowerment • Energy • Freedom • Instinct
Intuition • Power • Strength • Trust**

HORSE

THE HORSE IN NATURE AND CULTURE

Hanging in some of the highest peaks in the Himalayas are prayer flags adorned with the image of a horse. Known in Tibetan as *Lung ta*, or Windhorse, this spiritual symbol represents the speed, strength, and energy of the horse, combined with the wind's freedom to move unimpeded over the Earth. Within Tibetan Buddhism (as well as in older indigenous folk religions), the Windhorse carries prayers to heaven and delivers positive energy across the world. The horse clearly possessed qualities that created the impression that he, as opposed to any other animal, was capable of such a tremendous feat.

Tibetan Buddhists aren't the only ones who saw the horse in such a positive light. The Native American people of the Great Plains placed high value on the horse, which gave them a way to travel faster, hunt better, fight more fiercely, and carry more. The Lakota word for horse—*sunkawakan* (an anglicized spelling of šúŋkawakȟáŋ)—is composed of *sunka*, meaning dog, and *wakan*, meaning great or sacred. The name reflected the way the horse, like the dog, was loyal and helpful, only to a much greater extent. Horses were so strongly associated with empowerment that the Lakota and other Plains tribes described stealing horses as stealing power. The Mongolians have a similar relationship with horses, reflected in their expression "A Mongol without a horse is like a bird without the

As herd animals, horses naturally follow dominant individuals, whether horse or human.

wings." Throughout the world, horses have long been regarded as having so much strength and stamina that supernatural abilities were attributed to them, such as being able to travel between worlds or act as spirit guides.

The association of horses with power is nearly universal. Horses are large, strong animals capable of crushing a human being without much effort. And yet, even a child can climb onto the back of a horse and, with nothing more than the strength of her bond and a few relatively subtle gestures, convince the half-ton animal to move the way she wants him to. Once mounted and in command of such an undeniably majestic animal, a rider can move swiftly and cover great distances, while feeling an exhilarating sense of freedom and power.

However, what horses do for individual riders is nothing compared to what they have done for human civilization. As the saying goes, "History was written on the back of the horse." Between the horse's role in transportation, trade expansion, exploration, hunting, battle, and agriculture, he may have changed human history more than any other animal. The horse not only gives us the *feeling* of power, but has given us *actual* power to transcend human limitations in countless ways. The pervasive idea of the horse as a source of speed and strength is embodied in the term horsepower, which equates horses and engines based on the number of horses an engine could replace.

Although horses are prey animals who would rather flee than fight, they will stand their ground and aggressively defend themselves when necessary. All horses—and stallions in particular—can be very excitable and will sometimes stampede out of control. As a result of their

Through their connection with both worlds—the fenced and the free— horses provoke us to think about how we find a balance between the wild and tame parts of our own natures.

passionate natures, horses are associated with instinct, impulse, and sexuality. In the wild, a stallion often leads a small group of horses—known as a band—and enjoys mating privileges with multiple mares. For this reason, stallions came to symbolize sexual vitality, which gave rise to the "stallion in bed" idiom for a lustful, virile man.

Because horses are social animals, they form companionship attachments to one another, as well as to other species, including humans. As herd animals, horses naturally follow dominant individuals, whether horse or human. The horse's ability to be domesticated connects him to the concepts of trust and service.

INSIGHTS FROM THE HORSE

We hitch them to carts, carriages, and sleighs, we saddle them and bridle them, mount them and ride them. And they comply by granting us a nobility of service. Yet, despite their consent to be tamed and trained, horses remain more than a little wild at heart. They always have at least one hoof on the other side of the domestic divide. We can see it in them when they tear across the earth—the wind in their racing gait, a storm in their thunderous hooves. Through their connection with both worlds—the fenced and the free—horses provoke us to think about how we find a balance between the wild and tame parts of our own natures. Do we give free rein to our own wildness on occasion, letting ourselves break away from whatever constrains us?

Before a human can ride a horse, she needs to first learn how to read the horse's nuanced expressions, and to never forget that the horse is reading her the same way. A rider cannot deceive herself about how she feels when getting on a horse, because the horse can sense her emotions. In this way, horses teach us that before we can own and control any kind of power, we must know ourselves—our strengths *and* our weaknesses. If we do not know who we really are, we might get thrown, both literally when riding, and metaphorically whenever we are in a situation where

The horse's speed, strength, and energy inspired the idea of the Windhorse, a mythological Tibetan creature who carries prayers to heaven and delivers positive energy around the world.

self-knowledge is essential to our success and even safety. The horse, therefore, forces us to face our shadows—those aspects of our personalities that we do not consciously recognize or acknowledge.

Furthermore, because horses symbolize drive, they prompt us to think about what it is that motivates us and whether these urges and impulses sometimes cause us to stampede out of control. When passions get the better of us, can we let reason take the reins? Are we capable of holding our horses, of stopping to think before rushing into the fray? Or are we unbridled in our actions? In the same way that we can harness and ride a horse, we can harness our own vitality and power, but only if we understand the forces that propel us. Once we familiarize ourselves with these energies and gain control over them, we can use them to carry us further than we thought possible, while gracefully jumping the hurdles that life lines up for us.

Courage • Energy • Impossibility • Magic
Timelessness • Tirelessness • Vitality

HUMMINGBIRD

THE HUMMINGBIRD IN NATURE AND CULTURE

Writer and philosopher Sam Keen once wrote, "Forests are enchanted enough without elves or hobbits. Did you ever see a ruby-throated hummingbird?" For those of us who have seen hummingbirds, we know what Keen meant. There's almost nothing ordinary about these creatures.

To start with, they are nature's smallest bird, with one species—the bee hummingbird—measuring only two and a quarter inches in length and weighing less than a dime. Then there's their iridescent plumage that shimmers and shifts from one color to another depending upon the angle from which you view them. And if their size and coloring aren't magical enough, the way they fly will pretty much convince you that they crossed over from the fairy realm. Often described as the stunt pilots of the avian world, hummingbirds can rotate their wings—which beat as many as eighty times per second—in a circle. This enables them to fly forwards, backwards, sideways, up, down, and hover in place. They are so fast that sometimes all you see is a bejeweled blur.

All of these characteristics—from their light-scattering, color-changing feathers to their impressive acrobatic flight—have associated hummingbirds with the ability to accomplish that which seems

Considered the stunt pilots of the avian world, hummingbirds can rotate their wings—which beat up to eighty times per second—in a circle.

impossible. When they hover in mid-air, they create the illusion of stopping time, which has made them symbols of timelessness.

Inside the hummingbird's head is another surprise: the biggest brain in the animal kingdom, based on brain to body weight ratios. The hummingbird's brain represents 4.2 percent of his body weight, whereas the human brain represents only 2 percent of total body weight. Scientists are still learning about what the hummingbird does with this big brain, but one thing they have determined is that the hummingbird has an incredible memory. The hummingbird visits, on average, one thousand flowers a day and accurately remembers their locations, which flowers have the most nectar, and how recently he fed from each one.

So much activity—physical and cerebral—requires rest, and hummingbirds really know how to conserve energy when necessary. While sleeping, or when temperatures drop or food is in short supply, they enter a hibernation-like state called torpor, in which their heart rate and body temperature drop and their breathing slows dramatically. By entering torpor, hummingbirds can save up to 60 percent of their available energy. Certain hummingbird species go into the death-like state of torpor and then "return to life" once the day warms up, so they are symbolically connected to the concepts of rebirth and rejuvenation. For example, the Aztecs believed that fallen warriors were transformed into hummingbirds, and their god Huitzilopochtli—linked to the sun and warfare—was sometimes depicted as a hummingbird.

The hummingbird's colorful plumage links him to the sun, rain, and the rainbow, and by extension, fertility. The Hopi and Zuni peoples of the American Southwest paint hummingbirds on water jars as a talisman for bringing rain. This tradition is based on a myth about hummingbirds intervening on behalf of humans, convincing the gods to bring rain.

Hummingbirds compete for nectar and insects and are sometimes fiercely territorial. They have been known to chase off birds as large as jays and hawks, which is why they are associated with courage. Their fiery and passionate natures also find expression in their elaborate courtship rituals, during which males put on spectacular shows of aerial acrobatics. Given their courageous behavior and ardent courtship, it's no surprise that hummingbirds symbolize vitality, strength, and male virility.

Despite their showy courtship behavior, hummingbirds don't form bonded pairs of parents that share in the raising of their young. Instead, after mating, most male and female hummingbirds take up separate nests, with the male primarily taking on the role of guarding the local food source. They are a largely independent species who rarely travel or gather together, except at feeders. Even during their migrations—sometimes as long as four thousand miles—hummingbirds travel alone. As a result of their seemingly constant motion and incredibly lengthy migrations, hummingbirds symbolize stamina, endurance, and tirelessness.

The hummingbird visits, on average, one thousand flowers a day and accurately remembers their locations, which flowers have the most nectar, and how recently he fed from each one.

INSIGHTS FROM THE HUMMINGBIRD

When hummingbirds are in motion—hovering here, darting there, and then whizzing away in a blur—they seem impossibly energetic. As they helicopter from one nectar source to another, they don't waste any time. In fact, hummingbirds have been seen somersaulting from flower to flower, like aerial gymnasts. These indefatigable birds can't help but motivate us to get up, get moving, and go taste whatever sweetness the world has to offer.

The male hummingbird is no less energetic in his courtship. He performs showy dive displays in which he flies as high as one hundred and fifty feet up, then plummets straight down (at the speed of 385 body lengths *per second*), and then pulls up against the force of gravity at the final moment. If the macho dive bombing doesn't impress the female, he will try a gentler, more romantic approach—called a shuttle display—in which he flies back and forth in front of the female, gently serenading her and signaling his interest by expanding and contracting his tail feathers. He might even turn his head from side to side to show off his glittering plumage. The male does all this to win a mate, however it's ultimately the female's call, and sometimes his efforts end in rejection. But this doesn't seem to ruffle the male's dazzling feathers, or his self-esteem. He simply waits for the chance to do it all over again when he finds another female. He is unshakable, and offers us an inspiring model of iridescent resilience.

Clearly, the hummingbird works hard and plays hard, but he rests even harder. When not flying, feeding, or courting, he often enters the deep state of torpor to replenish his reserves. In doing so, the hummingbird helps us to remember that rigorous activity, as well as stressful conditions, require retreating from the world and recovering. If we want to live life like a hummingbird—somersaulting through the flowers and courting with passionate zeal—we need to take the time to refuel.

The hummingbird might be the smallest bird, but he lives life larger than many. He is courageous, passionate, and resilient—a bejeweled embodiment of spirit who teaches us to hover in the moment, find nourishment in beauty, and savor the nectar of life.

Elusiveness • Night • Perception • Power
Stealth • Transformation

JAGUAR

THE JAGUAR IN NATURE AND CULTURE

The word "jaguar" derives from *yaguára*, a Tupi-Guarani word that translates as "he who kills in one leap." As the third largest cat after the tiger and lion, and the strongest cat pound for pound, the jaguar lives up to his name. Capable of jumping twenty feet in a single bound, the jaguar can take down animals as large as cattle and as aggressive as crocodiles. With the strongest jaws of all the great cats, the jaguar also can crush bones, skulls, and even turtle shells with just a single bite. These large cats (weighing up to two hundred pounds) hunt the same prey as humans, have no predators other than humans, and have been known—on rare occasions—to hunt humans, too.

Even from a distance, the jaguar inspires fear and respect—but imagine how it feels to live in the same world as this formidable predator. The ancient indigenous peoples of Mesoamerica and South America were so awestruck by the jaguar that they regarded him as the "master of animals," and made him an emblem of power, authority, and military might. Mayan and Aztec kings sat on thrones covered in jaguar fur and used the name and image of the jaguar as a symbol of sovereignty; and an elite group of Aztec soldiers, known as Jaguar Warriors, wore uniforms inspired by the spotted pattern of jaguar pelts.

The ancient indigenous peoples of Mesoamerica and South America were so awestruck by the jaguar that they regarded him as the "master of animals," and made him a symbol of power, authority, and military might.

Not simply strong, but also stealthy, the jaguar excels at concealing himself, and is the most secretive of all the great cats. His striking spotted coat is surprisingly effective camouflage for his forested habitat, enabling him to blend into foliage and fade into patterns of shadow and light. The jaguar is so good at remaining invisible that experienced trackers have hiked in known jaguar habitat, concluded there weren't any cats in the area, and then discovered the cat's prints next to their own when backtracking on the trails. Even the scientists who study the jaguar rarely see this elusive cat.

When the jaguar does make an appearance, it could be anywhere. One might encounter a jaguar sleeping in a cave, crouched in the upper branches of a tree, or silently swimming across a river, his head barely visible above the water. To the ancient peoples who lived in jaguar territory, the cat's ability to occupy so many different habitats made him seem not only the master of animals, but the master of realms, too. These cultures attributed meaning to features of the land, such that caves were seen as portals to the spirit world and repositories of ancestral memory, and trees were thought of as transitional zones between heaven and earth. Because jaguars appeared in these places, people came to see them as creatures with access to special wisdom who could travel between worlds.

In addition to inhabiting a variety of landscapes, the jaguar also lives flexibly within the world of time. Although he is primarily nocturnal, his circadian rhythms can shift so that he is as comfortable hunting at dawn as he is at dusk, or even in the middle of the day. For people who share his territory, the jaguar's unpredictability is a given: they never know where or when they might come across this fearsome cat. However, encountering a jaguar at night is especially unnerving. Like other cats, the jaguar's eyes have a mirror-like structure that reflects light, causing them to glow eerily. His flashing eyes, combined with his superior night vision, contributed to the jaguar's reputation for possessing supernatural abilities—in this case, amplified perception.

Those who still live alongside the jaguar and retain their shamanic view of the world associate this cat's mercurial habits with shape-shifting. They believe that the jaguar can transform himself to look like a different animal, or even a human—and the reverse is also thought to be true: a shaman could take on the appearance of a jaguar. So one could never be sure what one was seeing—a cat or a person. In ancient times, the people of these cultures also believed that their gods sometimes took the form of jaguars. For example, the Aztec god Tezcatlipoca, whose name translates as "Lord of the Smoking Mirror," carried an obsidian mirror. He was said to use this mirror—associated with the jaguar's all-seeing eyes—to peer into the hearts of humans. This god had a jaguar counterpart, Tepeyollotl, who was considered to be both the god's companion as well as an aspect, or embodiment, of the god himself. Sometimes called the "Jaguar of the Night," Tepeyollotl helped Tezcatlipoca to see into the darkness of the world.

INSIGHTS FROM THE JAGUAR

As the quintessentially elusive cat, the jaguar calls our attention to our own abilities to evade notice, comprehension, or capture, reminding us that there are times when it serves us best to be inconspicuous. Not revealing everything about ourselves can sometimes give us distinct advantages, from avoiding challenging predicaments to simply lending us an air of mystery. The jaguar tells us that in order to have the edge, we need to keep them guessing.

The jaguar's mythical shape-shifting ability offers us two provocative concepts: personal transformation and symbolic embodiment. Even though we can't easily change our physical forms, we can change our hearts and minds by rewriting the narrative of our lives. We can "shift our shape" by striving to be more fluid in our self-expression, exploring new roles for ourselves, and expanding our comfort zones. Even just reframing our perceptions—in order to see the world differently—can

The jaguar's ability to occupy so many different habitats—from treetops to water—contributed to beliefs about his supernatural powers.

result in remarkable inner and outer changes. Second, the idea of transforming into an animal invites us to imagine what it is like to "become jaguar" (or any other animal). By doing so, we dissolve some of the perceived differences between humans and animals, and reconnect to our own animality, as well as to the commonalities we share with other creatures.

Finally, the jaguar prompts us to think about how we manage our fears. Sometimes it really is "a jungle out there," and our fear of the unknown can incapacitate us, limiting our options. But we can't always avoid the "jungle" just because a "jaguar" might be stalking us. We need to find a healthy balance between holding ourselves back in order to stay safe and taking risks in order to explore new opportunities. One of the ways to deal with fear is to become more familiar with the hidden parts of ourselves, known as the shadow: the emotions, thoughts, and other aspects of self that we have either rejected or suppressed. Our fears often take cover in this darker thicket of our psyches, but if we can shake them out of the underbrush and accept what we find there, we become less afraid. Then, with our fearless "inner jaguar" by our side, we can walk more boldly through the world.

**Cooperation • Dominion • Guardianship • Loyalty
Power • Sociability • Solar • Strength**

LION

THE LION IN NATURE AND CULTURE

Roughly half a million years ago, lions ruled the earth. They ranged over a territory larger than that of any other land mammal. Lions were everywhere: in Africa, Asia, Europe, and the Americas. They lived in locations now known as London, Los Angeles, and the Yukon. Back then, the only places you couldn't find a lion were Antarctica and Australia.

It seems fitting that the lion, so long associated with sovereignty, once had a nearly global kingdom. We don't know exactly when the lion first captured the interest of early humans, but radiocarbon dating tells us that our fascination with the large cat goes back to paleolithic times. The famous Chauvet cave paintings in France, created about thirty-two thousand years ago, include strikingly realistic images of lions. Archeologists believe that the artworks could only have been rendered so accurately if the artists had carefully studied their subjects. Even more ancient than Chauvet is the world's oldest-known zoomorphic sculpture, known as the "lion-man." Carved out of mammoth ivory—approximately forty thousand years ago—it stands just over a foot tall and depicts an upright figure with the head of a lion on a partly human body. Considered by many archeologists to be the oldest-known evidence for religious beliefs, the sculpture suggests that

Because of their social behavior and group hunting strategies, lions symbolize cooperation, fellowship, and loyalty.

the lion has played a role in our mythic imagination since the dawn of time.

Although we can only guess about what the lion signified in prehistory, it's likely that his position as an apex predator (capable of taking down large prey, such as wildebeests, gazelles, zebras, giraffes, and even elephants), along with his size, ferocity, strength, speed, and confident bearing made as strong an impression on the imagination of early humans, as it did on those who left historical records. What we do know, from some of the earliest art, myths, and spiritual traditions, is that the lion has long been a universal symbol of power, dominion, and majesty, which gave rise to the epithet "king of the jungle." Similarly, the term for a group of lions—known as a pride—is inspired by the lion's aura of dignified nobility and self-assurance.

The lion's roar—used to communicate as well as to mark and reinforce territory—further links him to the idea of invincible authority.

Although all four great cats—the lion, tiger, leopard, and jaguar—have specially adapted larynxes that enable them to roar, the lion has the most thunderous voice of all. Often as loud as 114 decibels (twenty-five times louder than a gas-powered lawn mower), a roaring lion can be heard from as far as five miles away. Reverberating through the air, the king of the jungle's roar sounds like a proclamation of sovereignty. In Buddhism, the Sanskrit word *Simhanada* translates as the "Lion's Roar," and refers to the intensity of the moment of enlightenment. More generally, roaring is used as a metaphor for empowerment, as in "I am woman, hear me roar," or a surge into a position of success, as in "the team roared back in the fourth quarter."

The lion's halo-like mane and golden fur also connect him to light and the life-giving, fiery energy of the sun. In Egyptian mythology, the warrior goddess Sekhmet ("She who is powerful")—depicted with the head of a lioness and crowned with a solar disk—was associated with

The lion's passionate nature inspired the African folk belief that sleeping on a lion skin will increase virility.

chaos, war, and the sun. Her fiery breath was the hot desert wind, and she had the ability to destroy or heal, punish or protect. The theme of protection appears in other lion lore and can be found in an old African folk belief that lions sleep with their eyes open, which links them to watchful guardianship. One of the best known leonine guardians is the Great Sphinx, the mythical creature with a lion's body and a pharaoh's head who guards the pyramid tombs at Giza in Egypt. The lion as a symbol of protection can still be found today in stone lion statues that guard entranceways.

Despite the persistent epithet "king of the jungle," the vast majority of lions live in open savannah and grassland habitats that have provided humans with the opportunity to watch them and learn about their behavior. Lions are the most social of all cats and form prides in which all of the females are related—as mothers, daughters, grandmothers, and sisters. The females usually stay together for life, cooperatively

Lions are the most social of all cats and form prides in which all of the females are related—as mothers, daughters, grandmothers, and sisters.

hunt together, collectively care for and defend one another's cubs, and allow for equal breeding opportunities with the males. The males primarily provide protection and a chance to breed, though they will hunt for their prides on occasion. They often form coalitions made up of two to six males—brothers, half-brothers, cousins, and occasionally unrelated males—who bond and work together to protect their prides and acquire new territory.

Members of a pride will sleep together in a pile of lush golden fur, their tawny bodies sensuously draped over one another as they snooze away the hottest hours of the day. When awake and not hunting or eating, lions snuggle, rub their heads together, and nuzzle and lick one another. Because of their social behavior and group hunting strategies, lions became symbols of cooperation, fellowship, and loyalty.

Lions have voracious appetites—for food, territory, and sex. Their tendency to claim the largest share of whatever they want inspired fables—generally ascribed to Aesop—in which the lion goes hunting with other, less exalted animals. When it comes time to divide the prey, the lion always proposes a plan that gives him either the largest portion or the entirety of the meal, which inspired the idiom "the lion's share." The moral of such stories is that business partnerships with "the mighty" are never a good idea. However, romantic partnerships might not be so bad. When it comes to sexual stamina, the lusty lion reigns supreme. When females are in estrus, lions will mate as many as one hundred times a day, for many days in a row. The lion's passionate nature inspired stories, notions, and metaphors, such as the African folk belief that sleeping on a lion skin will increase virility, and the German idiom for a ladies' man, *Salonlöwen* ("living-room lion").

INSIGHTS FROM THE LION

There is nothing subtle about the lion as he saunters across the savannah, exuding the confidence that justifiably belongs to a top predator secure in his kingdom. He knows what he wants and goes for it, as any sovereign would. These regal cats can't help but prompt us to think about our own lives and the extent to which we have dominion over them. Do we reign over our territories—whether personal or professional? Or do we too often let the proverbial hyenas steal our quarry? The lion reminds us to guard that which is near and dear to us.

Utterly unselfconscious, the lion inspires us to hold our heads high as we protect our pride—our loved ones *and* our sense of dignity and self-worth. The charismatic golden cat also invites us to let our own light shine—openly and proudly—so that we can be admired as the radiant creatures we are. His resounding roar further invites us to not be afraid to use our own voices to empower ourselves.

Reverberating through the air, the lion's roar sounds like a proclamation of sovereignty, which is why roaring is used as a metaphor for empowerment.

Lion prides teach us the value of working cooperatively in our families and communities. Female lions have even been called feminists because, being both accomplished hunters and devoted mothers, they "do it all." Furthermore, they are incredibly liberated in their egalitarian breeding habits and not shy about indulging their fiery passions. They also provide various kinds of support to one another. Not only are they known to help care for one another's cubs, but as a lioness ages and starts to suffer from physical infirmities that impact her ability to hunt, younger lionesses in her pride share their kills with her.

Despite their cooperative behavior, lions also call our attention to the dark side of power and its tendency to lead us to take ethical shortcuts. If we are awash in our own authority, we might become too domineering or take the lion's share of attention in our relationships or careers. Being overly prideful can also distort our perceptions of ourselves and others, leading us to behave arrogantly and foolishly.

Finally, the lion—like all large carnivores—invites us to examine the predatory energies in our own natures. Such instincts and urges can sometimes rush us into conflicts, causing us to devour those who get in our way. When we sense these impulses, most of us respond by denying or repressing this ferocious inner lion. But there is another, more lion-hearted approach: preserve safe habitat for these feelings in our psyches and let them roam free there, until we can calmly assimilate them into the totality of our expansive leonine selves.

Adaptability • Expression • Primal • Sensitivity • Survival

LIZARD

THE LIZARD IN NATURE AND CULTURE

When we imagine an iguana the size of an elephant, what appears in our minds looks like a creature from the Jurassic period. If we similarly envision a gigantic version of the Armadillo lizard—a species covered in a spiky coat of armor—a mythological dragon emerges in our mind's eye. This is what intrigues us about lizards: they have an anachronistic aura, as if they traveled here through time from an earlier era when dragons sparked the collective imagination or from the primeval past when dinosaurs ruled the earth.

There are nearly seven thousand species of lizard, ranging from the tiny Jaragua lizard of the Caribbean, who measures just over half an inch long, to the Komodo dragon of the Indonesian islands, who can grow to more than ten feet in length and weighs more than three hundred pounds. All lizards have certain characteristics in common, such as being cold-blooded (they get their heat from the environment), smelling with their tongues, scaly skin, a sinuous body and tail, and four clawed feet. However, many species have particularly unique and distinctive features.

Certain groups of lizards—such as chameleons, anoles, and some species of geckos—can change the color of their skin. Contrary to popular belief, scientists discovered that chameleons do not change colors

Lizards look like they have traveled here from the primeval past, when dinosaurs ruled the earth.

in order to blend into their environment, as their baseline coloring generally serves this purpose. Instead, the chameleon changes color to warm up or cool down (darker colors absorb more heat), to communicate mood, to announce sexual availability and impress potential mates during courtship, to establish sexual disinterest (as when a female is pregnant), or to warn enemies.

Male chameleons will often engage in "color battles" to show their rivals—through a coded display of hues and patterns—that they are more powerful. The weaker lizard will eventually "wave his white flag," and end his show of colors. As a result of these and other displays, such as inflating themselves or expanding frills, chameleons and other lizards are linked to the idea of somatic expression, or communicating through body language.

So when we describe someone as chameleon-like, meaning that they quickly change their appearance, opinions, or behavior to blend in, the facts do not support the metaphor. It would be more accurate to describe someone who reveals their mood through bodily expression—by blushing when embarrassed or turning pale when frightened—as being like a chameleon.

In addition to changing colors, lizards have developed a wide variety of other unique adaptations. Chameleons have tongues—as long as two and a half times their body length—that they can shoot out with lightning speed and marksman-like precision to catch insects. They also have prehensile tails that can wrap around objects when climbing and eyes that can move independently, enabling them to see in two directions at once. There's a folk saying in Madagascar—home to many species of chameleon—that advises, "Walk like a chameleon—look forward, but watch your back." The maxim was inspired not only by the chameleon's unique vision, but also by his slow and deliberate walk, which suggests an air of cautiousness.

Chameleons aren't the only lizards who have developed unique abilities. For example, geckos can climb vertically and even upside down using their unusually sticky toe pads. Komodo dragons developed a venom that kills their prey. If a Komodo gets a good bite in, he can walk away, wait for the prey to die, then return to devour it. Iguanas on the Galapagos Islands have adapted to swim in salt water, and the "scuba-diving lizard"—who lives near streams in Costa Rican mountains—creates an air bubble on his snout, which he then uses like a scuba tank so that he can breathe underwater.

The only place in the world where one might still see a soaring serpent is in South and Southeast Asia, which is home to the Draco lizard, also known as the Flying Dragon. This small lizard (rarely longer than eight inches) developed the ability to glide from tree to tree—sometimes as much as two hundred feet—using folds of skin that open into an umbrella-like set of wings when extended. Another amazing adaptation can be found in species such as the Australian frilled lizard and the

Male chameleons will often engage in "color battles" to show their rivals—through a display of hues and patterns—that they are more powerful.

South American basilisk, who can become bipedal and run upright on their hind legs. The basilisk can move so quickly on his hind legs that he can even run across the surface of water.

Like mythical dragons and extinct dinosaurs, many lizards have crests, frills, horns, spines, and other anatomical features that help them to camouflage or defend themselves. Several species inflate parts of their bodies when threatened to make themselves look larger. Some lizards have even evolved the ability to shed and regenerate their tails—a defense mechanism enabling them to escape predators. Once released, a tail sometimes writhes and squirms, which holds the predator's attention while the lizard makes his getaway. In ancient Greece and Egypt, lizards were regarded as symbols of good fortune because of their ability to regenerate their lost tails.

All of the different ways that lizards have evolved connect them to ideas of survival and adaptation. For several Native American cultures, the lizard, who could endure drought, thirst, and unrelenting heat, was associated with protection, healing, survival, and good luck. Plains Indian tribes sewed a newborn boy's umbilical cord into the shape of a lizard as an amulet of health and strength.

Lizards have acute senses and can detect even extremely subtle changes in their environment. Certain lizard species, such as some iguanas and monitors, have a third photoreceptive eye—known as a parietal eye—located at the top of their heads. This eye uses light from the sun to help these lizards find their way and to help moderate circadian rhythms and thermoregulation. The lizard's acuity and the way he is so tuned into the environment links him to sensitivity, intuition, and sensory intelligence.

INSIGHTS FROM THE LIZARD

When we consider all the unique physical attributes and behaviors that lizards have developed, it seems clear that the saurian way—the "Tao of lizards," so to speak—is about evolutionary adaptation. On a smaller time scale, our ability to adapt to changes in our personal lives is often the only way to live a life without unnecessary strife. The Tao of lizards, therefore, reminds us to find creative ways to accommodate life's challenges.

Now that scientists know lizards do not change their skin colors to blend in, but instead do so mostly to communicate, lizards have given us a new metaphor—this time for emotional transparency. The saurian way is to let one's true colors shine. There's no second-guessing what sort of mood these creatures are experiencing. They wear their hearts on their sleeves, and can inspire us to do the same when emotional authenticity would serve us better than pretense.

Chameleons can read one another loud and clear—they are either in the mood or not, ready to fight or not. We might not communicate through a code of colors and patterns, but we nonetheless express ourselves through posture, gestures, breathing patterns, facial expressions, blushing, teeth clenching, and more. Saurians remind us that

Lizards have developed countless fascinating adaptations, such as the sticky toe pads of the gecko, which enable him to climb vertically and even upside down.

our bodies talk, too, and the more attention we pay to these somatic expressions, the more information we gather, which has the potential to enhance communication.

The lizard's ability to lose his tail provides us with a powerful metaphor for cutting our losses—being able to recognize when it is better to walk away than it is to stay in the game. The lizard may lose his tail, but gets away with his life. For us, when we can let go of something that isn't working, we might lose an investment of time, resources, or even heart, but we gain the opportunity to start over.

The silent lizard basking in the warmth of the day seems so acutely tuned into his environment as to be practically merged with the world. He reminds us of the pleasures of losing ourselves in our senses—of lying on a warm stone, absorbing the heat from the sun, feeling the texture of the rock under our bodies. The lizard's peaceful, meditative presence reminds us to give into those moments when time seems to stand still and the boundaries between the inner and outer worlds start to dissolve until we feel at one with everything.

The lizard is so tuned into his environment that he has been described as having an "undifferentiated consciousness" in which he appears to become one with his surroundings.

**Abundance • Adaptability • Cunning • Elusiveness • Fertility
Foresight • Gnawing • Infestation • Smallness • Timidity**

MOUSE AND RAT

THE MOUSE AND RAT IN NATURE AND CULTURE

In Aesop's fable "The Lion and the Mouse," a mouse mistakenly awakens a lion sleeping in the forest. The lion captures the mouse, but before he swallows her, she begs for mercy, promising to someday repay him if he lets her live. The lion laughs at the ridiculous premise that a mouse could ever help him. However, feeling generous, he lets her go. A few days later, the lion finds himself trapped in a hunter's net, and his angry roars echo through the forest. The mouse recognizes the lion's voice, runs to him, chews through the ropes, and frees him. The fable ends with the rodent proudly boasting that even a tiny mouse can help a mighty lion.

"The Lion and the Mouse" is just one of many folktales in which the mouse, despite her size—and sometimes because of it—surprises everyone by saving the day. From Aesop's fables of ancient Greece to indigenous tales of the New World, mice symbolize not only smallness, but the potential for smallness to be powerful. In myths and stories, they often play the role of the prey who doesn't stand a chance against the predator—and yet prevails. In these tales, mice defy our expectations by succeeding against the odds. For example, in Native American Algonquian and Athabaskan stories, it's a mouse who saves the world by freeing the sun after it—like Aesop's lion—had been snared in a

In ancient times, the burrowing activity and nocturnal habits of rats and mice connected them to the underworld and cast them in the role of mediators between the living and the dead.

trap. The mouse's smallness doesn't limit the seemingly insignificant little creature, but instead elevates her.

Mice—and their closely related cousins, rats—belong to a group of animals known as Muridae, or murids, which includes more than 1,383 mouse-like species. Rats are larger than mice, and have smaller ears relative to the size of their heads, blunter snouts, and hairless, scaly tails. Although both rodents are intelligent, curious, and social animals, the rat is often painted as a bolder, scrappier version of her smaller cousin and so serves as a metaphor for sly, cunning, and selfish behavior. This is why when someone cheats, we call them "a dirty rat," and when someone snitches on another, we describe it as "ratting" on them. Mice, in contrast, are perceived as timid, more easily frightened, "quiet as a mouse" homebodies who live mousey domestic lives.

Historically, rats were blamed for spreading the plague—known as the "Black Death"—around Europe in the fourteenth century. This is why rats—and to a lesser extent mice—became symbols of pestilence. For years, historians and scientists speculated that the fleas on the rats spread the disease, but recent computer models suggest that the disease-carrying fleas jumped from human to human, not from rat to human. Nonetheless, there are many diseases that mice and rats can—and do—spread to humans, so their reputation as vectors is warranted. Their tendency to move into our dwellings, invade our pantries, and reproduce in our homes, along with their ability to spread disease, links mice and rats to the idea of infestation—on municipal, architectural, and microbial levels. Ironically, in the laboratory, mice and rats are regularly—and often very cruelly—sacrificed in the service of human health.

Both mice and rats are omnivorous and eat everything from nuts, berries, and bark to snails, insects, and whatever they can find in human pantries. In addition to using their teeth to eat, they also use them to gnaw through just about anything—walls, wood, electrical wires, and even certain kinds of concrete—to make passageways or manufacture material for nesting. The word "rodent" comes from the Latin verb *rodere*, meaning to gnaw or eat away, and this is what all rodents, including mice and rats, do: they chew constantly. They must do so in order to keep their teeth—which never stop growing—from getting so long that it would impact their ability to eat. Their constant gnawing serves as a metaphor for anxious states of relentless trepidation or worry, as when something is gnawing at us.

Mice and rats are found all over the world—on every continent, except Antarctica. The brown rat and the house mouse are considered the most successful mammals on earth, possibly even exceeding humans in their numbers. Mice can give birth to ten litters a year, with as many as fourteen pups per litter. Rats can give birth roughly

In many cultures around the world, mice symbolize not only smallness, but the potential for smallness to be powerful.

once a month, with up to twelve pups per litter. Without considering environmental factors (such as human extermination efforts), a breeding pair of rats or mice can generate up to fifteen thousand descendants in just one year. No wonder a group of these rodents is called a "mischief."

Because they can survive and thrive nearly anywhere, mice and rats are associated with adaptability and abundance. As one might expect of such prolific breeders, they are also linked to fertility—but not merely because of their own fecundity. Throughout history, rats and mice have ruined harvests by eating crops in fields and storehouses. Perceived as having the capacity to destroy a harvest (through eating) and to create one (through restraint from eating), they were thought to have control over the crops. Because people believed that they had such power over the harvest, they looked to mice for signs about what was to come. Mice then became associated with the idea of prophecy and inspired the ancient practice of myomancy—reading omens from the behavior or movements of rats or mice. The Haida, an indigenous people of the Pacific Northwest, practiced myomancy, as did ancient Greeks and other peoples.

Small and stealthy, the mouse and rat have long been associated with sneakiness and stealing. The English word for mouse comes from the Latin *mus*, meaning "thief." In the New World, in Náhuatl—the language spoken by the ancient Aztecs—the word for mouse was *quimichin*, a term used as a metaphor for spies. The invisibility of mice—the way they are present, but unseen—also link them to secrecy. In ancient times, the burrowing activity and nocturnal habits of mice and rats lent them a mysterious aura that connected them to the underworld and cast them in the role of mediators between the living and the dead.

The gods who were associated with mice and rats were often depicted as trying to protect the world from them. For example, the Hindu elephant-headed god, Ganesha, regarded as the remover of obstacles, was often shown riding on or attended by a mouse or rat. This symbolized his ability to keep their destructive tendencies under control, and to enter, like a mouse, any of the nooks and crannies where obstacles might be hiding. Similarly, the Greek god Apollo was sometimes described as Apollo Smintheus, Lord of Mice, and portrayed with a mouse lying at his feet or in his hand.

INSIGHTS FROM THE MOUSE AND RAT

The mouse and rat show us that big things can come in small packages. They are not large, but have nonetheless made a larger-than-life impression on the human imagination because of their capacity to grow in number, destroy crops, or eat through virtually any barrier. These tiny rodents remind us that bigger isn't always better and, more often than not, it's the little things in life that matter most. Just remember: two mice on day one, but a year later—as many as fifteen thousand mice.

Small and stealthy, the mouse and rat have long been associated with sneakiness and stealing. The English word for mouse comes from the Latin *mus*, meaning "thief."

Our perceptions of mice and rats represent two different, but familiar, ways of being. There are days when we feel quiet, timid, and more than a little mousey. We just want to retreat to our homes with a well-stocked larder and enjoy life's simpler pleasures. Other days we feel more like a rat: we want to hit the streets, see what's new to eat, and take our chances in the world.

Mice and rats are always burrowing, tunneling, chewing, and otherwise making their way into inaccessible spaces in search of food and nesting materials. This behavior serves as a metaphor for a certain kind of unrelenting, probing thought process that does not cease until we have found what we are looking for. Although this behavior often works for rodents, sometimes it doesn't, and they get caught in traps or chew on the wrong kind of wires. In this way, they also serve as a warning to proceed with caution when feeling driven.

Surpassed only by insects in their ability to make us feel invaded and infested, mice and rats call our attention to the need for good boundaries, and the value of recognizing signs suggesting that we have been breached. Just as we inspect the foundations of our homes, we need to periodically assess the vulnerability of our psyches to determine if we are sufficiently protected from infesting or parasitic thoughts. Similarly, the ceaseless nibbling of rats and mice reminds us that when something is gnawing at us—whether a hunger for something new or a persistent worry—we probably need to address it because it's unlikely to go away on its own.

Mice and rats are not likely to go away either. They have been on the planet for nearly seventy-five million years, whereas humans have only been around for about six million. As for which of us will inherit the earth, given their incredible resilience, it will most likely be the mice and rats.

**Change • Depth • Evolution • Flow • Involution
Order • Protection • Withdrawal**

NAUTILUS

THE NAUTILUS IN NATURE AND CULTURE

Nautiluses are ancient sea creatures who have survived five mass extinctions, including the Permian, two hundred and fifty million years ago, which wiped out roughly 90 percent of all life on Earth. While the planet suffered catastrophic changes and species came and went, the nautilus lingered on, hiding from predators deep below the surface of the sea. Not only did the nautilus survive all these extinctions, but her anatomical structure has remained largely unchanged throughout the eons. This enduring creature has had nearly the exact same form for the last five hundred million years, which is why they are often described as "living fossils."

The nautilus, like the octopus, squid, and cuttlefish, is a mollusk belonging to a class of animals known as cephalopods. Similar to other cephalopods, she is comprised primarily of a head and tentacles. But unlike any other living cephalopod, she has an external skeletal structure that takes the form of a chambered shell. As the shell grows, it wraps around itself, preserving the earlier stages of growth. The soft-bodied nautilus then moves into the larger chamber. The nautilus begins her life with four chambers, adding more as she grows. Over the course of their lives (which are long compared to other cephalopods, sometimes more than twenty years), nautiluses have been known to create up to thirty chambers in their shells.

When the nautilus's shell becomes too small, she creates a new chamber to move into—one that still protects her while accommodating her growth.

The spiral form of the nautilus's shell connects her to the symbolism of this archetypal shape, which has long fascinated humankind, no doubt because matter and energy tend to constellate in spirals. Galaxies, storms, the horns of mountain sheep, the proportional unfolding of rose petals, the rhythm of the human heart's bioelectrical impulses, and the descending flight path of a falcon to her prey—all of these forms, motions, and energies have spiral patterns that expand outward from or inward to a fixed point.

The spiral—and by extension, the nautilus—represents the creative power of the universe and the concept of constant change, as well as evolution and involution. Because the nautilus's shell is a logarithmic spiral and maintains the same proportions throughout her life, it also symbolizes the idea of unified, ordered growth. For this reason, the Greeks regarded the nautilus's shell as a symbol of perfection. As for the cephalopod herself, she can withdraw completely into her shell and close the opening with a hood. Like other animals who retreat in such a manner, the nautilus is associated with self-protection, withdrawal, and shyness.

Nautiluses have a highly developed sense of smell and as many as ninety tentacles that they use to taste and touch. They are both predators and scavengers, and use their tentacles to grab onto prey and pull it toward their mouths. Their eyes are primitive and lack a true lens, probably because these creatures spend so much time in the darkest depths of the sea. The nautilus uses jet propulsion—by drawing water in and out of her shell's chambers—to swim. The nautilus regulates her buoyancy by adjusting the amount of water and gases within each of her shell's chambers. During the day, the nautilus descends—sometimes as much as twenty-five hundred feet below sea level—but at night, she slowly rises to feed in shallower waters.

INSIGHTS FROM THE NAUTILUS

The nautilus's shell, like the tides, mirrors the flow outward and inward, and reminds us of our own cyclical urges to sometimes fully engage the world and other times go within when we need shelter, solitude, or renewal.

The nautilus's shell grows in such a way that it protects the soft animal within without impeding her growth. When her shell becomes too small, she creates a new chamber to move into—one that still protects her while accommodating her ongoing development. We would do well to be inspired by the nautilus's strategy: making room for our personal growth while still finding ways to shelter ourselves.

The unfurling of the nautilus's shell—the way it grows by building upon its earlier stages—offers us a metaphor for the developing

During the day, the nautilus descends—sometimes as much as twenty-five hundred feet below sea level—but at night, she slowly rises to feed in shallower waters.

self. Each of us starts out as a "one-room house" and gradually, over the course of our lives, we build addition upon addition with experience after experience. Our past slowly wraps around us, becoming the multi-chambered shell, or self, we live within. This self—the spiraled, storied life we build—carries us forward into the uncertainty of the future while simultaneously comforting us with the familiarity of the past. In this way, the nautilus reminds us that although we are wrapped in the past, we are not stuck in it: the future is always ours to shape.

Because she can retreat into her shell, the nautilus connects us to our own periodic urges to withdraw from the world and rest quietly within the self.

**Adaptability • Change • Depth • Elusiveness • Mystery
Otherness • Shape-shifting**

OCTOPUS

THE OCTOPUS IN NATURE AND CULTURE

Octopus researchers often say that if we want to imagine what extraterrestrial intelligence might look like, we need look no further than the octopus. About five hundred million years ago, the octopus evolved from a simple snail-like ancestor. Over time, this squishy, soft-bodied creature became what some scientists have described as the first intelligent life on Earth. With eight limbs that attach to her head, three hearts that pump blue blood, and brain cells in her arms, the octopus's anatomy is undeniably otherworldly.

The extraordinary octopus has inspired both ancient seafaring storytellers, who wrote of the monsters of the depths of sea, and contemporary science fiction writers, who feature her as an alien intelligence from the far reaches of outer space. The most famous mythical octopus-like creature is the *kraken* of Scandinavian folklore. Not only could this frightening monster grab ships and sailors and pull them down into the depths; she could also create powerful ship-sinking whirlpools.

Ancient Hawaiian myth, however, casts the octopus in a different light. The creation story begins by explaining that the current universe is merely the remnant of a much older one that left behind just one survivor—the octopus, the only creature who was capable of slipping through a thin crack between the two universes. In 2018, a group

Ancient Hawaiian myth tells us the current universe is the remnant of a much older one that left behind just one survivor—the only creature who was capable of slipping through a thin crack between the two universes: the octopus.

of thirty-three scientists seemed to echo this ancient Hawaiian myth when they (seriously) theorized that the octopus didn't originate on Earth but came here more than five hundred million years ago on a comet carrying fertilized octopus eggs that crashed into the sea.

In addition to being the poster child for otherness, the octopus has an impressive array of talents that gave rise to her other symbolic associations. To start with, the octopus is a master escape artist. In one of his books, Jacques Cousteau relates a story about how one of his friends placed an octopus in an aquarium and covered the aquarium's top with a weighted lid to prevent the octopus's escape. Not long afterward, his friend found the octopus in his library, exploring his books and turning the pages with her arms. There are countless other reports of octopuses escaping from tanks in labs and aquaria. Sometimes, when they get free of their own enclosures, they crawl across the floor to another tank, slip inside, and snack on unsuspecting fish and crustaceans. Other times, they make a break for the outside, leaving only suction-cup tracks as a clue to their escape route.

Octopuses are good at escaping because they are shape-shifting contortionists who can practically liquify themselves as they squeeze their soft, boneless bodies through tiny holes and crevasses. But their elasticity is simply one trick up their eight sleeves. Another is to change the color, pattern, and texture of their skin to blend into their surroundings or to mimic the appearance of another species. Divers have reported seeing one species—the mimic octopus—use this camouflaging ability

With eight limbs that attach to her head, three hearts that pump blue blood, and brain cells in her arms, the octopus's anatomy is undeniably otherworldly.

to impersonate a wide variety of animals, from anemones and jellyfish to seahorses and stingrays. And even more impressive than their mimicry skills is their ability to know which creature they should impersonate based on the local who-eats-whom ecology. For example, when scientists observed a mimic being attacked by damselfishes, they saw it imitate a banded sea snake, a known predator of damselfish. Finally, when threatened, an octopus will sometimes release a cloud of dark ink as it swims to safety. With so many ways to change her form and slip away, no wonder the octopus is associated with illusion, inconstancy, and elusiveness.

If an octopus loses an arm, no problem—she can regenerate it, which is convenient because she has learned to make the most of her many arms. When using her arms as "legs," this spidery cephalopod can climb out of the water, walk on land, and even run—sometimes moving as quickly as a cat. When using her arms like "hands," the octopus demonstrates impressive dexterity: she can pass a ball from sucker to sucker, open clams and childproof pill jars, build stone walls, and untie knots. The octopus's dexterity not only contributed to her portrayal as a grasping monster of the deep, but also made her a symbol of lust and lechery because she is always reaching out and touching that which interests her.

INSIGHTS FROM THE OCTOPUS

As a soft, shell-less cephalopod, the octopus is relatively defenseless, so she developed other ways of protecting herself, demonstrating that there are many ways to avoid confrontation if we are flexible. Taking a page from the octopus playbook, we might elude notice by blending in with our surroundings or changing our presentation of self to appear more like another. If all else fails, we could create a metaphorical cloud of black ink—by obfuscating a situation—and slip away while others try to puzzle it out. At the same time, the clever, shape-shifting octopus also suggests that we question appearances and not be deceived by the illusions that others create.

With her many arms and suckers, the octopus can remind us of those times when we feel pulled under or entangled, suggesting that it may be time to disengage, rise to the surface, and catch our breath. Conversely, the octopus also presents the ultimate example of being able to let go. Because of her ability to regenerate a limb, the octopus can—metaphorically speaking—walk away from a situation, even if it means losing an arm. There are times when we hold on too tightly, when what we actually need to do is to let go, even if it means temporarily losing a part of ourselves.

More than anything else, the octopus invites us to transcend the barriers of difference; to see beyond the obvious and to be open to that which is unlike us. In the 2016 science fiction film *Arrival*, an intelligent, tentacled, ink-squirting, octopus-like alien race arrives on Earth. At first, humans fear these creatures, seeing them as a threat and potential enemy. By the end of the film, however, humankind succeeds in opening itself to the aliens' otherness, gaining benefits that ultimately help expand human consciousness and save humankind. In our own lives, we can reap many rewards by moving beyond our comfort zones, widening our perceptions, and expanding our circle of relationships beyond those who look or think like us.

Agility • Courage • Curiosity • Dexterity • Friendship
Nurturing • Playfulness • Trust • Vivacity

OTTER

THE OTTER IN NATURE AND CULTURE

The Ainu, an indigenous people who live on Hokkaido (Japan's north-ernmost island), have a creation myth that begins with God getting interrupted before he was finished creating humankind. He is unex-pectedly called back to heaven to deal with important business, and so arranges for a lesser deity to step in and finish the task. Prior to his departure, God summons an otter and asks him to deliver to the deity specific instructions on how to complete humankind. The otter prom-ises to do so, but then spends his time amusing himself by fishing and frolicking and consequently forgets to deliver God's message, causing humankind to be imperfect and not exactly as the Creator intended.

Even though this myth tells us that we could blame our flawed human nature on the otter's carefree nature, who would want to? Just the mere sight of otters playing lifts our spirits. A family of roughhous-ing otters is the very embodiment of joyful exuberance, which is why the collective noun for a group of otters is a "romp." When playing on land, otters wrestle, tumble, chase one another, and slide down slippery rocks and riverbanks, often squeaking and squealing the entire time, which only adds to the impression that they are having a blast. In the water, they dive, twist, and twirl together with remarkable agility and gracefulness. When not engaged in these athletic frolics, otters explore

A family of roughhousing otters is the very embodiment of joyful exuberance, which is why the collective noun for a group of otters is a "romp."

the world with a puckish curiosity, often playing with sticks, stones, and other objects. They even juggle: they toss and roll pebbles between their forepaws and around their heads and necks as dexterously as a seasoned street performer. Given their energetic antics, it's no surprise that otters became the animal kingdom's ambassadors of play.

With their streamlined bodies, smooth fur, webbed feet, and rudder-like tails, otters are designed for diving and swimming. They can close their nostrils and ears and stay underwater for up to eight minutes. Otters appear to pour themselves into the water, streaming through their aquatic environments with such supple fluidity that they seem to merge with the current itself. In the water, otters are strong, swift hunters whose whiskers pick up changes in the current that indicate the presence of predator or prey. As a result of all their aquatic adaptations, otters are linked to water and its symbolic associations: emotions, intuition, imagination, and receptivity.

Because the otter moves between land and water, and because she dives down to the depths and then emerges at the surface, she was perceived as a mediator between realms or states of being, including the transition from illness to health. For the Ojibwa—an indigenous people of North America—the otter is sacred and plays an important role in their curing rituals. In one of their traditional stories, the Great Rabbit (the servant of the Great Spirit) summoned Otter and gave him the sacred drum, rattle, and tobacco as tools to cure the sick. The Great Rabbit also conveyed to Otter the secrets and mysteries of curing and made him immortal, which is why Ojibwa healers use otter pelts for the medicine bags that contain their sacred curing items.

Otters are very non-confrontational and are largely able to avoid aggression through an elaborate scent-marking system used to establish boundaries. During times of plenty, males will look the other way when intruders briefly cross into their territories. Females, despite

By following the old maxim that "good fences make good neighbors," otters remind us that we all fare better with good boundaries.

being especially protective and attentive mothers, will allow their off-spring to play with pups from other mothers. Their relatively peaceful management of territoriality, devoted maternal care, and relaxed mingling of unrelated offspring connect the otter to the concepts of nurturing and trust.

Despite their relatively peaceful natures, otters will fiercely defend their young if necessary and have earned a reputation for being incredibly brave. The Cree, an indigenous people of North America, tell a story about a time when the sun was captured and tied up with rope. Otter was the only one brave enough to attempt a rescue. While working to cut the ropes, Otter's fur was burnt off, his eyes damaged, and his teeth reduced to stumps . . . but Otter never gave up. After the sun was freed, He-Who-Made-the-Animals rewarded Otter by giving him all the wonderful qualities he has today, from his thick, insulating coat and unrelenting courage, to his agility and jaunty, joyful spirit. The otter's ability to be both ferociously brave and charmingly carefree at the same time has been recognized and admired by many cultures.

One of the most distinctive characteristics of sea otters is their use of stones to break open oysters and other mollusks, as well as crustaceans and sea urchins. Many otters have what appears to be a favorite rock that they carry with them wherever they go, in a "pocket" of loose skin under their forearm. The sea otter's tool use links her to dexterity. To stay warm in the water, sea otters have two layers of incredibly dense fur—the thickest in the animal kingdom—with more than a million fibers per square inch. The air trapped in their fur keeps them warm and gives them buoyancy, which is why sea otters can spend hours floating on their backs, often in large groups known as "rafts." These communities of floating otters, who sometimes hold hands, have made sea otters symbols of sociability, friendship, and affection.

INSIGHTS FROM THE OTTER

Above all else, the otter inspires us to lighten up, seize the day, and have some fun. Otters continue to amuse themselves even as adults, reminding us that doing something simply for the fun of it need not be a trait relegated to the realm of childhood. Despite the otter's playful nature, she can be a formidable and fearsome creature if those near and dear to her are in danger. In being able to balance both sides of her nature—fun-loving and fierce—the otter shows us that disparate aspects of personality can be gracefully and healthily integrated.

The otter's agility offers us a metaphor for the ability to move easily, both physically and mentally. As the antithesis of rigidity, otters remind us to keep our bodies and minds flexible so that we can more easily adjust to our ever-changing lives and the challenges that await us at each new phase. The otter's movement through water further evokes the idea of fluidity, and of being able to go with the flow and deal with situations without trying to assert unnecessary control. Likewise, her association with receptivity reminds us to stay open to ideas and possibilities.

Finally, by following the old maxim that "good fences make good neighbors," otters remind us that we all fare better with good boundaries, whether those mark property, personal space, or other kinds of territories. They demonstrate that establishing rules and following them results in conflict avoidance, which ultimately fosters trust and peace. This state of calm, in turn, gives us more time to slide down riverbanks, float on our backs while holding hands, or otherwise enjoy the world as our oyster.

**Darkness • Invisibility • Magic • Mystery • Night • Perception
Prophecy • Secrecy • Wisdom**

OWL

THE OWL IN NATURE AND CULTURE

As dusk dissolves into darkness, an eerie call cuts through the last breath of light. Suddenly, seemingly out of nowhere, an owl emerges, her eyes flashing as she silently descends on an unsuspecting mouse. Seizing the rodent in her sharp talons before he even senses her presence, she and her quarry vanish into the night.

It's easy to understand why, after witnessing such an unnerving spectacle of nocturnal predation, early peoples would come to associate owls with supernatural powers—their actual abilities are as awe-inspiring as those of any mythical creature. To start with, owls can see and hunt in the dark with a precision that would be impressive in daylight, let alone at night. Their huge eyes are densely packed with retinal rods—the photoreceptors that enable sight in dim light. The front-facing position of their eyes provides them with binocular vision, which increases depth perception and improves focus. The result of these enhanced visual abilities is that owls can see up to one hundred times better than humans can in low light.

Owls also have incredible hearing. The feathers on their facial disks collect and funnel sounds into their ears, which are located beneath their eyes. Because they can direct so much sound into their ears, and because their ears are designed to precisely discern the direction from

The owl's night vision can serve as a metaphor for seeing into those dimly lit places in ourselves where we hold secrets or keep ourselves "in the dark" about something.

which a sound is coming, owls can locate prey moving under foliage or snow by sound alone. Further, owls can swivel their heads up to 270 degrees, enabling them to direct their visual and auditory focus to wherever they detect sound or movement.

As a result of their forward-facing eyes, inscrutable stare, and ability to see and hear things we cannot—particularly the way they can sense prey from a distance in advance of its appearance—owls have long symbolized astute perception and wisdom. The best known example of the owl's association with wisdom comes from ancient Greece, where the owl was sacred to Athena, the goddess of wisdom. Wise owls also appear in the myths and fairy tales of many cultures, and find expression in the contemporary owl-as-teacher motif.

Ancient peoples not only believed that owls were wise and possessed superior vision; they also imagined that owls had supernatural abilities, such as the power of foresight and the ability to reveal secrets hidden in the human heart. The owl's perceived powers and spooky nocturnal habits played into humanity's universal fear of the dark and inspired the widespread belief that owls could predict death, disease, or destruction. Even today, in parts of rural Europe and throughout many African cultures, there is a pervasive belief that owls are harbingers of doom. Similarly, in numerous Native American cultures, to see, hear, or even dream about an owl is an unlucky omen.

One rarely sees or hears an owl until the last minute. This is because owls have specialized feathers that enable them to fly silently until they are within inches of their prey. When air rushes over the wings of other birds, it creates turbulence, which makes a whooshing sound. In contrast, an owl's wing feathers reduce turbulence and noise, especially when descending at the steep angles of flight required in the final seconds before striking prey. So when one encounters an owl in flight, she

The owl's nearly silent flight gives her an incredible advantage when hunting prey.

often seems to have emerged out of thin air, which led many cultures to regard the owl as a spirit guide or messenger capable of moving between worlds. This is why Inuit shamans wear the feathers of the snowy owl to help them ritually move between the realms of the living and the ancestors.

INSIGHTS FROM THE OWL

Despite the owl's reputation for having extraordinary sight in the dark, the owl's real visionary power is in being able to see at dawn and dusk—those in-between times of ambiguity and uncertainty when shadows and light play tricks on us. The owl's superior vision during these liminal times invites us to think about how we perceive—and try to resolve—ambiguity in our lives. Do we let our senses adjust to the dusky light before we try to interpret the landscape? Or do we jump to conclusions before we can make out the shapes in the shadows? The owl can inspire us to navigate liminality by patiently

As a result of their forward-facing eyes, inscrutable stare, and ability to see and hear things we cannot—particularly the way they can sense prey from a distance in advance of its appearance—owls symbolize astute perception and wisdom.

watching and waiting for resolution, instead of descending on the wrong targets.

As a bird known for her nocturnal vision, the owl calls our attention to darkness—both outer and inner. Her night vision can serve as a metaphor for seeing into those dimly lit places in ourselves where we hold secrets, or keep ourselves "in the dark" about something. The owl, therefore, reminds us to turn our gaze inward—as well as outward—when seeking answers or clarity.

Sitting on her perch, the watchful owl offers a lesson in paying attention with all our senses. As a result of how carefully she attends to the world, she is capable of detecting even the most subtle changes in her environment. The leaf litter under a tree flutters just the slightest bit, and she tilts her head toward it to listen. She hears a faint stirring, and then sight and sound combine to create an image in her mind—a mouse hiding under the leaves. The owl swoops in, and seconds later she has swallowed her supper. The wise raptor shows us that the simple but profound act of paying attention merges the world with the mind in a way that can reveal that which is hidden.

**Community • Fidelity • Humor • Intelligence • Love
Mimicry • Mischief**

PARROT

THE PARROT IN NATURE AND CULTURE

Alex—an African grey parrot who was the focus of a famous animal cognition study—loved being tickled around his neck. He would often approach someone, bend his head to expose the back of his neck, and say, "You tickle." Pretty much everyone obliged and tickled him. Once, a few days after receiving a toy parrot as a gift, Alex approached it, bent his head, and uttered his "You tickle" request. When the toy parrot failed to tickle him, Alex looked at it and said, "You turkey!" No one had taught Alex to use the word "turkey" as a reprimand for noncompliance. He had figured it out by himself, from research assistants who often called him a turkey when he didn't respond the way he was supposed to.

As special as Alex was, he is far from the only parrot who has demonstrated an ability to associate human words with their meanings and to use them correctly and creatively. Many parrots have impressed scientists, as well as their human companions, with their capacity to use words and to engage abstract concepts—such as shape, color, and number—in ways that suggest true comprehension. But long before parrots were the subject of cognition studies, they were widely known for their intelligence, mimicry skills, playfulness, and mischief, and had come to symbolize these attributes.

Long before parrots were the subject of cognition studies, they were widely known for their intelligence, mimicry skills, playfulness, and mischief, and symbolized these attributes.

Found in warm climates all over the world, there are more than 350 known species of parrot that vary in size from the tiny, three-inch-long buff-faced pygmy parrot to the bright blue hyacinth macaw, who can grow to forty inches in length. All parrots share two primary anatomical features that distinguish them from other bird families: blunt, rounded beaks and zygodactyl feet (with two toes facing forward and two facing backward) that are ideal for climbing and swinging. Their colorful plumage, with its dazzling array of hues and patterns, is used to attract mates and as a means of camouflage.

Humans have kept and bred parrots as companions, status symbols, and for ceremonial purposes for more than three thousand years.

Their intelligence, playfulness, capacity to form emotional bonds, and rainbow-colored feathers appealed to cultures around the world. Although parrots first appear in European historical records in 327 BCE, when Alexander the Great brought tamed ring-necked parakeets to Greece, they were kept as pets in South America as far back as five thousand years ago, and in India three thousand years ago. Toward the end of the first century CE, the Pueblo peoples of the North American Southwest acquired scarlet macaws from Mesoamericans and bred them. Fascinated by the macaw's mimicry abilities and long life spans, the Puebloans used their feathers in ceremonial attire and as ornamentation on sacred objects.

Although there are a few other birds who can learn to produce the sounds of human speech, no other bird does so with as much proficiency as the parrot. Historically, the parrot's ability to mimic human language has evoked a variety of conflicting associations. These loquacious birds were seen as "truth tellers" who reliably repeated only what they heard, but were also linked to gossip and indiscretion for the same reason. They were perceived as intelligent for being able to learn human words, but they also represented the idea of not thinking for oneself, as in the verb "to parrot," which means "to repeat by rote."

In the wild, as well as in captivity, parrots are playful birds who appear to have a sense of humor. The kea, a large, intelligent parrot native to the alpine regions of New Zealand, is known as the "naughty alpine parrot" because of his fearless curiosity and often destructive antics. Keas pull trim off of cars and rummage through backpacks, among other activities. Scientists recently learned that keas, when playing in this manner often produce laughter-like calls that appear to inspire other keas to engage in similar kinds of mischief, much the way human laughter can be contagious.

Social and gregarious, parrots form flocks that forage together and roost in large groups at night, behaviors that link them to fellowship

Parrots are known to form profound attachments to their companions, regardless of whether they are birds or humans.

and community. Using vocalizations and body language, they communicate to one another about mood, food locations, predator threats, and other matters. Parrots are steadfast mates and caring flock members, and will not leave one of their own when ill or injured. When another parrot isn't present, parrots transfer this bonding behavior to their human caregivers, becoming protective and devoted companions.

Like people, parrots can become so attached to their human companions that they develop a jealous possessiveness, sometimes even attacking other household members perceived as threats. Because of their loyalty to one another, and their deep attachment to their chosen partner (or human companion), parrots symbolize love and fidelity. In myths and folktales, they were given the job of delivering romantic messages or cast as the companions of deities of love. Within the Hindu pantheon, Kamadeva, the god of love and desire, is often portrayed riding a parrot.

INSIGHTS FROM THE PARROT

Parrots tend to repeat exactly what they hear, often to others' detriment: they have snitched on cheating spouses, tattled on disobedient children, reported on the misdeeds of dogs, and engaged in often cringe-worthy repetition of profanity. In this respect, we are cautioned by the parrot's example to be mindful because unfiltered utterances or intemperate words can come back to haunt us. The parrot's perceived lack of discretion also teaches us to think before we repeat information we have heard secondhand, lest we cause harm or spread misinformation. While the parrot's mimicry skills remind us that imitation has been described as the highest form of flattery, mindless repetition rarely serves us well.

Because we talk to parrots and they talk back to us, often repeating our words, they have been humorously compared to therapists who

Parrots are perceived as intelligent because they can learn human words, but they have also represented the idea of not thinking for oneself, as reflected in the verb "to parrot," which means "to repeat by rote."

echo or paraphrase the client's words back to her. Within the therapeutic relationship, the goal for such repetition is to help the client feel heard and understood. Within the human-parrot relationship, the parrot's mimicry often elicits vaguely similar feelings in humans. When the parrot repeats our words, he, too, is looking for understanding. In the wild, a parrot learns vocalizations from his family and flock and then uses these calls to communicate. In captivity, a parrot learns the vocalizations of his human "flock" for the same purpose. Parrots therefore remind us that among social creatures—including humans—communication is crucial, the need to be heard and understood is deep and profound, and the bonds of fidelity we form through words foster lifelong friendships and loves.

One of humankind's most ancient dreams is being able to talk to animals—it reaches back to shamanic traditions, myths, and fables, and it lingers on in storybooks and films. The parrot's ability to learn human words connects us to this desire to reach across the species divide. In this way, the parrot invites us to extend our interest beyond our social and cultural spheres, and to open dialogue and seek understanding with those different from us.

Humans have kept and bred parrots as companions, status symbols, and for ceremonial purposes for more than three thousand years.

**Beauty • Confidence • Display • Nobility • Ostentation
Pride • Vanity**

PEACOCK

THE PEACOCK IN NATURE AND CULTURE

In an 1860 letter to a colleague, Charles Darwin wrote, "The sight of a feather in a peacock's tail, whenever I gaze at it, makes me sick!" The peacock—the male peafowl—is a large bird in the pheasant family known for his spectacular tail comprised of multi-colored iridescent feathers. Darwin was perplexed by how evolution could allow for something as cumbersome and seemingly useless as the peacock's gigantic tail. After all, the tail clearly had no adaptive value—it doesn't help him fly, fight, or forage, nor does it provide camouflage.

But, as it turns out, the peacock's extravagant tail does have a purpose: it attracts peahens, the female peafowl. The peacock uses his conspicuous fan of tail feathers, known as a train, in courtship rituals in which peahens select their mates according to the size, color, and other attributes of the male's tail display. If sufficiently smitten with a particular male and his tail, the peahen will mate with him, thereby passing on his genes. The peacock's tail turned out to be one of the most persuasive inspirations for Darwin's theory of sexual selection, which proposes that certain traits exist simply because they give the individual an advantage in the mating game.

Thousands of years before Darwin speculated that it was the peahen's love of the peacock's tail that preserved and perpetuated this

The peacock's tail turned out to be one of the most persuasive inspirations for Darwin's theory of sexual selection, which proposes that certain traits exist simply because they give the individual an edge in the mating game.

flamboyant trait, the peacock was already a symbol of beauty, pride, and vanity. Not only is the peacock's train striking in itself, but the way he displays it makes it all the more irresistible. He raises and arches his train into a fan that spans across his back and extends to the ground on both sides of his body. Standing there, surrounded by a breathtaking halo of color and pattern reminiscent of the heavens, the peacock looks like a celestial being. No wonder all the peahens go crazy. Then there's the peacock's confident strut, which seems to exude pride and self-esteem. However, because peacocks repeatedly display for their harem, they can also appear pompous and ostentatious, and have become emblems of these qualities, too.

The peacock's shimmering colors and the vaulted shape of his open tail feathers have been compared to the rising and descending rays of light at dawn and dusk, and the colorful eye spots on the ends of his tail feathers have suggested supernatural watchfulness. In Greek mythology, Hera, wife of Zeus and queen of the ancient pantheon, memorialized her servant Argus—a giant with one hundred eyes—by placing his eyes on the tail of the peacock. Similarly, within the Christian tradition, the tail feather eyes symbolized the all-seeing God, and sometimes the Church.

In addition to the symbolism arising from their tails and courtship displays, peacocks are also associated with rain because they often dance prior to a rainfall. Consequently, people believed that peacocks could predict or even manifest rain. In reality, the peacock originated in places that have monsoon seasons which coincide with the peafowl's breeding season, so their rain dances were most likely well-timed coincidences.

Some ancient peoples believed that peacocks could digest poisonous plants and snakes and not only survive, but somehow turn the poison into their beautiful tail feathers. This belief connected peacocks to the idea of transmutation—the ability to change something from one state of being into another. For similar reasons, Buddhists associated the peacock with the concept of the *bodhisattva*, a person who can transform the poison of ignorance into the jeweled colors of enlightenment that opens like the peacock's tail.

INSIGHTS FROM THE PEACOCK

Never shy about strutting his stuff, the peacock calls our attention to how we handle our own assets—whether physical features, talents, or charms. Modesty, a lack of self-esteem, or even cultural preferences for humility can leave us feeling uncertain about how and when to reveal, let alone revel in, that which makes us attractive. The peacock, however, lacks neither confidence nor clarity when it comes to using his gifts to further his goals. He's the walking embodiment of the maxim, "If you've got it, flaunt it." He doesn't shrink from displaying his

The white peacock—a variation of the blue peacock—is associated with purity, eternity, and unconditional love.

spectacular train because reticence isn't an option for the peacock if he wants a chance to mate and pass on his genes. The peacock, therefore, reminds us that "nothing ventured, nothing gained."

Conversely, the peacock also has been viewed as a negative symbol representing vanity, pretention, pomposity, and superficial display. Too much flamboyance can reflect an excess of ego, often associated with insecurity. Such ostentation can cast a shadow on what would otherwise be beautiful. In this way, the peacock cautions us to not be so caught up in our own need to be admired that we distract from or obscure our best qualities.

Finally, the peacock offers us an intriguing way to think about beauty. In his book *Survival of the Beautiful: Art, Science, and Evolution,* philosopher and musician David Rothenberg wrote, "Perhaps the most important thing about desire is that it generates beauty, something that lasts longer than any want." The peacock's awe-inspiring train of iridescent jewel-toned feathers exists because peahens *desire* it. We don't know if they find it beautiful—as that cannot be discerned from their behavior—but they certainly find it desirable, as it drives their choice in mates.

Together, the peacock and peahen demonstrate that beauty exists as the result of a dance between receptivity and radiance—being receptive to beauty energizes its radiance and the radiance of beauty energizes its reception. The peahen and peacock show us that beauty is in the beheld *because* it is in the eye of the beholder. For peafowl, the desire for beauty results in males with spectacular tails having the chance to mate, pass on their genes, and create more peacocks with spectacular tails. For humans, the desire for beauty plays a role in mate selection, too, but it also plays a role in poetry, music, the preservation of a rainforest, and so much more. These magnificent birds—he with his incredible tail and she with her love of it—invite us to indulge our desire for beauty: to seek it out, appreciate it, and be inspired by it.

Abundance • Excess • Fertility • Gluttony • Humility
Impurity • Uncleanliness • Voracity

PIG

THE PIG IN NATURE AND CULTURE

Humans invented the pig. Actually, it was more of a loose "collaboration" between wild boars and hunter-gatherers who decided to establish settlements. Roughly ten thousand years ago, in what is now Turkey, humans built villages and began producing garbage. It wasn't long before wild boars showed up to forage on the trash heaps. Soon after the scavenging boars appeared, people began to breed the gentler, tamer boars and raise them as livestock. Over time their ears flopped and their tails curled, they lost their camouflaging spots and stripes, their tusks vanished, and eventually they became the animals we call domestic pigs.

Other livestock animals are raised for their ability to work, such as pulling plows, or for their production of milk, eggs, or wool. Pigs, however, have always been raised pretty much for just their flesh. These ample creatures provide meat in abundance by producing numerous fast-growing offspring. As a result of their generative natures, pigs became associated with fertility, abundance, and wealth. In ancient Egypt, the pig was sacred to Isis, the goddess of fertility and agriculture, and women who wanted to conceive sometimes wore amulets bearing the image of a sow and her piglets. In Norse mythology, the boar was sacred to Freyja, the goddess of love and fertility, who was

Pigs have more than twenty different sounds that they produce for different situations, ranging from vocalizations piglets use to call for their mothers to those that alert other pigs to danger.

sometimes depicted riding a boar with golden bristles or with a boar as her companion.

Domestic pigs are remarkably intelligent, sensitive, and curious animals. Like dogs, pigs respond to their names and can be trained to come, sit, and stay. They can learn the names of other animals and objects, and can be taught tricks such as fetching a ball or jumping on command. Pigs are known to be consummate escape artists and have figured out how to unlock and untie nearly every kind of gate closure using their surprisingly strong and sensitive snouts. These clever creatures also are capable of using tools. For example, scientists have taught pigs how to use mirrors to find hidden objects, how to use joysticks (with their snouts) to move images on computer screens onto targets in exchange for treats, and how to adjust thermostats to regulate the temperature in barns. The pigs excelled at all three tasks.

Pigs' social and emotional lives are no less impressive than their intellects. They have more than twenty different sounds for different situations, ranging from vocalizations piglets use to call for their mothers to those that alert other pigs to danger. To produce what has been described as a porcine lullaby, mother pigs softly and rhythmically grunt to their piglets while nursing them. Pigs enjoy playing together and develop close bonds with one another, sometimes even appearing to grieve the loss of mates, family members, and companions. Pigs also are empathetic and responsive to the emotional states of other pigs.

Despite all these wonderful qualities, pigs have long been associated with many negative characteristics. As omnivores, pigs eat pretty much everything, from vegetables, fruits, nuts, and roots to earthworms, small vertebrates, garbage, and on occasion carrion and feces. Their tendency to consume substances considered unsavory led various cultures to demonize them and declare them unclean. Judaism, Islam, and other religious traditions throughout history have designated pigs as forbidden flesh, resulting in pigs becoming symbols of impurity.

Pigs also tend to eat more than they need to survive, so they have been perceived as greedy and gluttonous, which is why we describe the over-indulgent as "piggy" and the act of over-eating as "pigging out." Pigs also came to represent uncleanliness because they wallow in mud as a way to stay cool, prevent sunburn, and repel insects. The word pig is a multifaceted derogatory epithet in many languages. In English alone, "pig" is used as an insult for everything from lasciviousness and laziness to slovenliness and corpulence.

INSIGHTS FROM THE PIG

Close to the earth, both literally as a creature who roots in the soil for food and symbolically as an animal who indulges her sensuality, the pig reminds us of our own sometimes insatiable physical desires—our urges to indulge them, as well as the need to temper them. As an

Wild boars are smart and sociable—two traits shared by their domesticated cousins.

opportunistic omnivore, the pig can usually find a source of nourishment in the world, highlighting the value of flexibility. And yet, the pig's lack of dietary discernment inspired most of the pig's negative associations, so the pig also serves as a reminder to exercise restraint when it comes to our appetites and curiosities.

The true nature of the pig has always been obscured by what humankind has projected onto her. The human appetite—for just about everything—surpasses that of all other creatures, yet we push our own avarice into the shadows while using the word "pig" as a synonym for gluttony. Our species pollutes air, water, and soil in our endless consumption of resources, yet we associate pigs with impurity and filth. We have made the pig a symbol of so many of our own loathsome qualities and behaviors. In this way, the pig calls our attention to how we project undesirable qualities onto others—animals as well as humans—while neglecting to recognize those same attributes in ourselves.

Winston Churchill once remarked, "I'm fond of pigs. Dogs look up to you, cats look down on you. Give me a pig—he just looks you in the eye and treats you as an equal." Sadly, we cannot say the same about how we have treated pigs. With the industrialization of agriculture, pigs have been increasingly subjected to unconscionable factory farming practices. No other food production sector does as much harm to the animal it raises as the pork industry. It isn't a coincidence that the animal who is treated so poorly is also the one who is so obscured by our projections. The pig shows us that when we can't see others—animals or humans—for who they truly are, but only for what we project onto them, we are capable of treating them in ways we never would if we could see their true natures.

As for the pig's true nature: if given the chance to live a healthy life, the pig is a sensitive, smart, sociable animal who reminds us to enjoy life's pleasures. Contrary to the Biblical expression "Do not cast your pearls before swine," which advises one not to waste anything of value on those incapable of recognizing its worth, pigs seem quite capable of appreciating life's "pearls." In her book *The Good Good Pig: The Extraordinary Life of Christopher Hogwood*, naturalist Sy Montgomery wrote about a pig she rescued. Montgomery described Christopher as enjoying company, good food, belly rubs, the warm summer sun, and, seemingly, life itself. She summed up the true essence of piggishness when she wrote, "Christopher Hogwood knew how to relish the juicy savor of this fragrant, abundant, sweet, green world." When we let pigs be pigs, they teach us to celebrate our own delightfully piggish tendencies to indulge in all that life offers us, and to wallow in pleasure whenever we can.

**Cleverness • Fertility • Ingenuity • Luck • Magic
Sensitivity • Swiftness • Timidity • Trickster**

RABBIT AND HARE

THE RABBIT AND HARE IN NATURE AND CULTURE

According to an old folk belief (still popular in Britain and parts of North America), saying "rabbit, rabbit" immediately upon waking on the first day of a month will bring you luck for the next thirty days, provided that the words are spoken aloud and before anything else is said. The reason the rabbit—as opposed to any other animal—is used for such an incantation is the rabbit's long association with luck, which arises from their fertility and wily ways.

Rabbits and hares are perceived as paradoxical creatures—courageous, yet timid; clever, yet foolish; and innocent, yet sexually prolific, which is why they have been portrayed as tricksters in cultures around the world. Wherever they appear, these long-eared characters usually outwit bigger, stronger creatures, often manipulating them to achieve their own goals. They break rules, defy authority, and inspire others to think outside of established norms. Br'er Rabbit, for example, originated in the folktales of enslaved Africans, in which he often undermined the plantation regime with his guile. Another African rabbit trickster—Kalulu from Zambian folklore—is so quick-witted and resourceful that his name came to mean "cleverness." In North America, among the Algonquin-speaking tribes of the northeast, the Great Hare, also known as Nanabozho, Mishabooz, and other names,

Rabbits and hares are courageous, yet timid; clever, yet foolish; and innocent, yet sexually prolific, which is why they have been portrayed as tricksters in cultures around the world.

is both a trickster and a culture hero who causes trouble in some stories, but saves the day in others.

Rabbits and hares differ in some significant ways. Even so, the terms are often used interchangeably (and sometimes incorrectly) in folklore and even in common names. For example, the jackrabbit is actually a hare and the Belgian hare is, in fact, a rabbit bred to look like a hare. As for the differences between the two, hares are larger than rabbits, have longer ears and legs, and run faster. Newborn hares, called leverets, are born fully furred with open eyes, and are ready to live on their own within hours of birth. In contrast, newborn rabbits, called kittens, are born furless, blind, and require their mothers' care for about eight weeks. Hares tend to be solitary, whereas rabbits tend to be more social, with some species living in stable communities comprised of family groups. The gentler, more timid rabbit has been domesticated; the wilder, less docile hare has not.

With eyes positioned on each side of their heads, rabbits and hares have nearly 360-degree panoramic vision, allowing them to detect predators from almost any direction. With long, sound-capturing ears that rotate up to 270 degrees, they hear sounds from greater distances and at higher frequencies than humans can. Their sense of smell—twenty times stronger than a human's—often enables them to detect predators before the predators can see them. Because of their acute senses and how tuned into their surroundings they are, rabbits and hares are associated with sensitivity.

Always ready to respond to threats with a repertoire of getaway strategies, rabbits and hares are notoriously difficult to catch. Sometimes they jump into action (as much as twenty-two feet in a single bound) and bolt away at lightning speeds (up to 25 to 50 mph). Other times, they run in a confusing zig-zagging manner—known as jinking—that confounds and eludes predators. The brown hare, for example, doubles back, leaps sideways, and performs other acrobatic tricks to create a perplexing maze-like scent trail that leads predators in circles. Hares will even jump into water and swim to evade capture.

Other evasive strategies include slipping into hidden underground passageways or camouflaging themselves in vegetation. When running, jumping, swimming, or hiding doesn't seem like the best strategy, they will sometimes simply freeze in place, as the lack of movement can make them difficult to spot. More often than not, hares and rabbits stay one step ahead of their predators, outwitting them through fleetness, ingenuity, cunning, and guile—qualities that they came to symbolize. But even when they fail to escape, their fecundity safeguards their species' survival.

Hares and rabbits are notoriously prolific, and the females of some species produce multiple litters and can give birth to up to fifty offspring a year. Female hares and rabbits will continue to mate, even when pregnant, and hares can even conceive twice and simultaneously carry two

One minute the rabbit or hare is visible, perhaps even dancing in a meadow under the light of the moon, and the next he is gone, prompting us to wonder if he was ever really there.

separate litters at different stages of development. Their enthusiasm for breeding, as well as their rate of reproduction, made them symbols of vitality and fertility, and linked them to abundance, luck, and spring-time. These associations gave rise to the mythical Germanic egg-laying Easter hare called *Osterhase*, and the more recent American spin-off, the Easter Bunny.

As a result of their passion for procreation, hares and rabbits became symbols of love, lust, and libido, hence the saying "to (procreate) like rabbits." Similarly, the idiom "mad as a March hare" is derived from the frenetic behavior of European hares during their spring breeding season. Their fecundity so impressed early cultures that, in ancient Egypt, the hieroglyph meaning "to be" is a long-eared rabbit above a stylized wave, and Osiris, the Egyptian god of fertility and rebirth (among other things), was sometimes depicted with a hare's head. In ancient Greece, the rabbit was sacred to Aphrodite, the goddess of love and fertility, and lovers were known to give each other gifts of rabbits as tokens of their affections.

Rabbits and hares are mostly nocturnal, so we usually see them at dawn and dusk. In the soft light of these liminal hours, our eyes can play tricks on us. One minute a rabbit or hare is visible, perhaps even dancing in a meadow under the light of the moon, and the next he is gone, prompting us to wonder if he was ever really there. For these reasons, rabbits and hares are linked to the mystery of the borderlands, passageways to other worlds, shape-shifting, and magic.

INSIGHTS FROM THE RABBIT AND HARE

Without their exquisite senses, the largely defenseless rabbit and hare could not survive. Always listening, watchful, and oriented, these sensitive and perceptive creatures show us the value of paying attention and bringing all of our sensory awareness to a situation. Doing so not only alerts us to threats, but also opportunities.

With their quick-witted survival instincts, agility of movement, and passionate drive to breed, rabbits and hares inspire us to embrace life with creativity, flexibility, and zeal. They remind us that there is almost always a way to outsmart our competition and those who challenge us, starting with the ability to think on our feet and move swiftly when faced with danger. Sometimes life gives us all the time in the world to think something over, but at other times we cannot afford to hesitate. Safety, or an opportunity, might vanish as quickly as a rabbit at dusk, so we need to hop to it. And if someone's on our trail—literally or metaphorically—and we cannot run fast enough, we can always double back, dive under, jink, or pull another trick out of our hat.

On the flip side, rabbits and hares also caution us about the negative aspects of an overly timid outlook. Their tendencies to be hypervigilant, remain close to home, hide, or freeze in place invite us to think

Rabbits remind us that there is almost always a way to outsmart our competition and those who challenge us, starting with the ability to think on our feet and move swiftly when faced with danger.

about how fear can limit us, depriving us of the richness of life. In the Native American story "Rabbit and Eye Walker," a rabbit fails to overcome his fears, loses a friend, and is then forever known as "Fear Caller" because of his tendency to dwell in fear. The moral of the story is that fear itself can be dangerous, sometimes summoning that which harms us. Life occasionally requires going outside our comfort zones, taking risks, and not staying stuck in timidity and trepidation. But even when we do get stuck, the rabbit and hare remind us that our courage—like the moon and the next cycle of fertility—will come full again. Rabbit, rabbit.

The rabbit's fecundity so impressed ancient Egyptians that their hieroglyph meaning "to be" is a long-eared rabbit above a stylized wave.

**Creativity • Light • Mischief • Playfulness
Transformation • Trickster • Wisdom**

RAVEN AND CROW

THE RAVEN AND CROW IN NATURE AND CULTURE

Raven was bored. And then something profoundly creative happened. Or destructive. Or both. In a story from the Haida people of the Pacific Northwest, the newly created world was beautiful, yet empty, and Raven was growing very tired of his own company. Then, just as Raven was flying over the coast, feeling miserably lonely, he heard a distant whimper and looked down. There, on the shore, was a giant clamshell, and hiding inside it were people. Raven wanted company, so he flew down and tried to coax the people out of their shell, but they were afraid. So, Raven talked to them as only Raven can—in his raspy, seductive voice. And it worked—the people left their shell and entered the world.

Among indigenous peoples, particularly those of the Pacific Northwest, and the Circumpolar North, Raven is often portrayed as a creator god, culture hero (a figure who embodies and often transforms a culture), and trickster. These traditional tales usually involve Raven finding a way to stir up some excitement—by creating the world; stealing light, water, or fire; or inventing something incredibly useful, such as language or the fish hook. Sometimes he causes trouble for his creations, just to amuse himself. Often, he is a shape-shifter, changing into human, animal, and inanimate forms at will. The raven's portrayal

Among indigenous peoples, particularly those of the Pacific Northwest and the Circumpolar North, Raven is often portrayed as a trickster.

as a trickster, along with his actual behavior in nature, has made him a symbol of creativity and transformation.

In many of the old myths about Raven, the story often begins with him not having enough to do. But it's not only mythology that has cast the raven as suffering from ennui. Scientists have similarly suggested that the playful antics of ravens—and their close relatives, crows—might arise out of boredom. It's possible that these bright, evolutionarily successful birds simply have too much spare time to spend on amusing themselves.

Consider some of the ways these birds play. In the Rocky Mountains, ravens figured out how to use curved pieces of bark to windsurf updrafts. Gripping the bark in their talons, they spread their wings and launch themselves into the air, using their feet to adjust the angle of their "surfboards" as they ride the wind currents. In Russia, a crow was captured on video using a jar lid as a makeshift snowboard to slide down a snowy rooftop. Both species play aerial tag, keep away, king of the mountain, and other games. Lying on their backs, they juggle objects between their feet and beaks. With such a repertoire of frivolous behavior, it's no wonder that ravens and crows became emblems of playfulness.

Ravens and crows are corvids—the avian family that includes magpies, jays, rooks, jackdaws, and other species. Ravens are considerably larger than crows, with bigger bills relative to their bodies. Their tail feathers are wedge-shaped (the crow's tail is fan-shaped), and have shaggier throat feathers, called hackles. The crow's call sounds like a caw, while the raven has a croakier voice.

Both birds are extremely intelligent. They can mimic the sounds of other animals, such as dogs and chickens, and inanimate objects such as sirens and car engines. They are also capable of learning to vocalize words and using them in appropriate contexts when communicating with people—a talent that undoubtedly contributed to their portrayal as shape-shifters. These birds can even use tools. One species, the New Caledonian crow, fabricates a tool that looks like a crochet hook and another that resembles a tiny saw, and uses both to hunt for insects. In Japan, when a traffic light stops cars, Carrion crows place walnuts on the street, wait for cars to drive over them and crack open the shells, and then return to eat the walnut meat. Hooded crows in Finland have been seen pulling on baited lines left in place by ice fishermen. The crows continued to tug on the line until they caught a fish.

Given such behaviors, it's not surprising that these birds have long been associated with wisdom. In Nordic mythology, the god Odin had two ravens—Hugin (Thought) and Munin (Memory)—who soared over the world, gathering information and bringing it back to him. In Welsh mythology, Bran the Blessed—whose name means "raven"—was a gigantic king whose enchanted severed head continued to provide wisdom, guidance, and entertainment long after it was removed from

Scientists believe that ravens and crows are so smart and successful as a species that they might simply have too much spare time to spend on amusing themselves.

his body. Bran's head is said to be buried under the Tower of London, and legend claims that as long as the ravens—believed to embody his spirit—continue to roost in the Tower, Britain will be protected.

In addition to being smart and playful, crows and ravens are social birds who travel in flocks when young, mate for life as adults, share parenting duties, and often live in extended families. They cooperate with one another, form alliances, share resources, and watch out for their young. Nesting adults will sometimes drop rocks from trees onto humans to keep them from getting too close to their eggs or nestlings. Larger groups sometimes engage in "mobbing," in which they gather together to dive at a perceived threat.

Being opportunistic omnivores, ravens and crows will eat almost anything, from seeds, nuts, and fruits, to insects, small mammals, garbage, and carrion. They will sometimes bury food in the earth so that it attracts maggots, which they later feast on—along with what's left of the food originally cached—thereby getting significantly more bang for their buck. Ravens and crows will also steal eggs and nestlings from other birds, as well as prey from predators. An otter about to eat a recently caught fish might find herself the victim of a classic crow heist in which two crows team up, with one crow pulling the otter's tail—causing her to turn around and drop the fish—while the other dives down to steal it. This sort of thievery and prankster-like behavior made crows and ravens symbols of mischief and contributed to their reputations as tricksters.

The crow and raven's dark plumage, carrion consumption, and menacing presence at gallows, graveyards, and battlefields caused European cultures to associate them with death, the devil, witches, and the otherworld. Some saw them as the souls of the damned or emissaries of Satan; others regarded them as omens whose appearance near a house foretold the death of an occupant. Through their black coloring they are symbolically linked to the night, gestation, and the magical ability to bring things into being. They are also widely associated with prophecy, a belief captured in the old Irish saying "to have a raven's knowledge," meaning to have the ability to foresee the future. The call of the raven, which often sounds like *cras*, is similar to the sound of the Latin word for "tomorrow," which is why the Romans associated the raven's call with hope and the idea of everything being better tomorrow.

INSIGHTS FROM THE RAVEN AND CROW

Whether stealing fish from otters or using car tires to crack open walnuts, ravens and crows are unabashedly opportunistic and readily take advantage of favorable circumstances. These roguish birds know how to take care of themselves and can inspire us to broaden our perspectives and engage a strategic mindset when it comes to tending to our own needs and trying to optimize our outcomes.

Crows and ravens are social birds who travel in flocks when young, mate for life as adults, share parenting duties, and often live in extended families.

With their incredible range of amusing antics, ravens and crows remind us to be more creative about how we entertain ourselves. Taking inspiration from these improvisational birds, we should periodically move out of the well-traveled lanes of our usual diversions, engage our spontaneity and ingenuity, and try something new. Research has shown that play is actually good for us: it relieves stress and improves cognition and memory (interestingly, Odin's ravens were named Thought and Memory). So, ravens and crows invite us to think of play not only as a way to have fun, but also as a way to sharpen our wits.

Despite their playful behavior, ravens and crows were taken seriously by ancient peoples. They were nearly universally perceived as possessing secrets—as knowing things that we do not. Even today, these inscrutable dark-winged birds connect us to mystery and to the idea of potentiality. They remind us that all things begin in the amorphous and chaotic gestational night, in which anything is possible.

Consider another story from the indigenous peoples of the Pacific Northwest. In the Tsimshian version, the world begins in darkness, and Raven is worried about how he will feed himself if he can't see well. Looking for a solution, he flies through a hole in the sky to heaven, where the Chief of Heaven keeps all the light locked up in a box. Raven notices the chief's daughter scooping water from a stream and turns himself into a cedar needle floating in her bucket. The chief's daughter swallows Raven, becomes pregnant, and gives birth to him in the form of a baby boy. As the boy grows, he charms his grandfather, who eventually allows him to play with the box that contains the light. The boy, of course, steals the light, turns back into his original form as Raven, and flies away, releasing the light into the world.

Raven shows us that finding and releasing the light isn't always easy. It often requires facing a challenge, going through a profound transformation, and giving things time to gestate. But, eventually, from chaos and night come creativity and light. This is Raven's gift to us.

Fatherhood • Love • Loyalty • Luck • Magic • Reversal • Wonder

SEAHORSE

THE SEAHORSE IN NATURE AND CULTURE

Everything about the seahorse is extraordinary, starting with the ways in which he appears to be something other than what he actually is: a fish. Instead of a streamlined horizontal body with scales, the seahorse has a vertical form with a gracefully curved neck and long snout, like a horse. His body is protected by an exoskeleton comprised of hard plates, like a crab, but covered in skin. His prehensile tail, like a monkey's, can grasp and hold onto objects, and his chameleon-like eyes can move independently, enabling him to see forward and backward at the same time. Similar to the octopus, the seahorse can change the color of his skin to blend into his environment or to express his mood. This unusual little fish even shares a characteristic with tigers—he growls when stressed. Finally, like a kangaroo, he has a pouch for carrying his young.

The male, however, not only carries his young, but actually gestates and gives birth to them. The male's role in reproduction makes the seahorse unlike any other known creature on Earth, and explains why the seahorse is associated with magic—he completely defies categorization and inverts the normal order of things. Because he was seen as magical, seahorse effigies, as well as dried seahorses, were used as good luck talismans.

The male seahorse not only carries his young, but actually gestates and gives birth to them, which makes the seahorse unlike any other known creature on Earth.

Ever the devoted mate, seahorses are monogamous throughout their breeding seasons and sometimes even for life. Seahorse courtship begins when a female finds a male that catches her eye. She visits him every morning, and the couple engages in courtship rituals: they dance together—circling each other, entwining tails, twirling and spiraling—all while changing colors. These morning dances, which can last from minutes to hours, solidify the relationship and synchronize their reproductive cycles.

When the big day finally arrives, the bonded pair approach each other, connect by touching snouts and bellies—their bodies forming the shape of a heart—and the female deposits her eggs into the male's brood pouch, where he fertilizes them. The male carries the eggs in his pouch for ten to twenty-five days, until they are ready to be pushed out into the sea by contractions in his body. By the time they are born, the young, called "fry," are fully formed miniature seahorses ready to live on their own. On average, a seahorse gives birth to 150 young at one time, but have been known to deliver as many as two thousand fry at once. However, on average, only five out of one thousand will survive to adulthood.

The seahorses' monogamous lifestyle and courtship dances have made them symbols of love and devotion in many cultures. Their unusual reversal of parental roles associates seahorses with fatherhood, and also links them to the ideas of inversion and departure from the status quo.

Because of his equine appearance, the seahorse inspired the mythological half-horse-half-fish creatures known as Hippocampus (horse-sea-monster), who pulled Poseidon's and Neptune's underwater chariots in Greek and Roman mythology. Within these ancient cultures, the seahorse was also thought to guide the souls of the drowned to the next world. It's also possible that the seahorse inspired the kelpie, the mythological shape-shifting Scottish water horse, and the *havhest*, a

Because of his equine appearance, the seahorse inspired the mythological half-horse-half-fish creatures known as Hippocampus (horse-sea-monster), who pulled Poseidon's and Neptune's underwater chariots in Greek and Roman mythology.

creature from Scandinavian folklore with the head of a horse and the tail of a fish. Seahorses usually are not found in such northern waters, but reports and specimens from afar could have contributed to the legends about these fabled beasts.

Shy, elusive, and secretive, the seahorse survives by largely staying in one place, blending in, and having an unpalatable exoskeleton. The male is a homebody, spending most of his time in a tiny territory—sometimes only half a square meter. Most of the time he is stationary, swaying in the currents with his tail wrapped around a blade of grass. When he does move, he is one of the slowest swimmers of all the fish in the sea. He swims in an awkward upright position, propelling himself with a small fin on his back that, despite fluttering up to thirty-five times per second, doesn't get him anywhere fast. Females are just as slow, but they move around more and across larger territories (roughly one hundred square meters).

But what he lacks in speed, the seahorse makes up for in stability and stealth. Despite his relatively docile demeanor, the seahorse is, in fact, a formidable predator. He can sneak up on prey and get very close without being seen or making a noticeable ripple in the water. Seahorses eat thousands of tiny shrimp per day, and have a 90 percent predatory success rate. The tiger, in comparison, has only a 10 percent success rate.

INSIGHTS FROM THE SEAHORSE

When a color-changing male fish who looks like a tiny dragon gives birth to as many as two thousand fully formed young, we can't help but feel a sense of awe and wonder. There's nothing about the seahorse that fits neatly into our rational understanding of the world. He defies our expectations and, in doing so, inspires us to be curious, which enriches our engagement with the world.

The seahorse reminds us that we can break out of personal or societal prescriptions and patterns, whether in traditional parenting roles

The leafy seadragon—who belongs to the same family of fish as the seahorse—also gestates and carries his young.

or other arenas. He invites us to gain perspective on other ways of doing things. His paternal pregnancy, in particular, provokes us to think about gender roles at large, including the way we balance traditionally defined masculine and feminine energies in ourselves, as well as in our families, workplaces, and society at large.

Because they are not strong swimmers and are often carried by sea currents, the seahorse frequently finds himself going with the flow, which can remind us that sometimes the best strategy is to simply yield to life's currents. Conversely, he also demonstrates the value of knowing how to anchor ourselves and hold on during tumultuous times.

The ability to simultaneously see forward and backward not only enhances the seahorse's hunting prowess, but better enables him to escape predators. In this way, he calls our attention to the value of maintaining broad perspectives that balance both foresight and hindsight, learning from the past while looking ahead, and keeping ourselves safe while pursuing opportunities.

The seahorse's slow and steady predation style—nine times more successful than that of the tiger—reminds us that we don't always need to be fierce, fast, or ferocious to get what we need. Patience and focus can be equally effective (and sometimes even more so) when it comes to accomplishing our goals.

As for the seahorse's love life: we will probably never know what seahorses feel during courtship, but their slow seductive dances, during which they flush different colors like blushing lovers, projects an irresistible image of idealized romance. And when the two bodies of a mated pair form a heart as they seal the deal, these magical creatures can't help but remind us of the mysterious drive that inspires us to seek wholeness through connection.

Balance • Imagination • Intuition • Playfulness
Shape-shifting • Transformation • Unconscious

SEAL AND SEA LION

THE SEAL AND SEA LION IN NATURE AND CULTURE

A widespread Scottish folktale often begins with a man fishing or walking along the shore where he encounters a group of female selkies—seals with the ability to shape-shift into humans. The selkies, who have shed their seal skins and are in their human form, are breathtakingly beautiful. They might be dancing in the moonlight on a beach or draped on rocks, quietly singing a haunting song. The man is immediately consumed with desire, so he steals a selkie's pelt and coerces her into becoming his wife. Without her magical skin, she is trapped in her human form, so she reluctantly agrees, but only for a limited time (often seven years). Throughout her time with her human husband, the selkie longs to return to her life as a seal, and as time passes without her seal skin, she starts to wither. Eventually, the selkie is reunited with her skin, shape-shifts back into a seal, and returns to the sea.

Found in Scotland, Ireland, and Scandinavia, stories about shape-shifting seals most frequently follow the pattern described above. Sometimes, however, instead of being captured by humans, selkies seek out their company. For example, there are tales of strikingly handsome male selkies who seek out women in unhappy marriages and seduce them. In other tales, the selkies act as good Samaritans by leading fishermen to promising fishing grounds or keeping swimmers safe.

Seals and sea lions are known to play games of chase and keep away, ride the surf, perform acrobatic stunts, and use other animals—such as starfish and pufferfish—as toys.

For those of us who have watched seals and their close relatives, sea lions, it's easy to see how these sea mammals inspired the idea of selkies. Their large, soulful eyes, friendly curiosity about humans, and ability to live both in water and on land made them the perfect inspiration for mythical shape-shifting creatures. Furthermore, there are many stories throughout history about positive encounters between these sea mammals and humans, which is why they have been described as the "dogs of the sea." Both seals and sea lions are known to seek out human company: they check out swimmers and divers; climb on top of kayaks, paddle boards, and surf boards; jump into boats; and engage water skiers. There are even reports of seals and sea lions assisting drowning swimmers and injured dogs who were struggling in the water.

Seals and sea lions are pinnipeds, a group of carnivorous flipper-footed mammals who adapted to an amphibious marine existence.

They use their webbed flippers to propel their streamlined bodies through water and to move about on land. They sleep on land as well as in water, where they float in a vertical position with their heads above the surface. They mate, give birth, and care for their young on land (or ice floes), but otherwise spend the rest of their time foraging or playing in their aquatic habitat. Sea lions and fur seals (a specific group of seals more closely related to sea lions) have visible ear flaps, longer flippers, and can walk, though somewhat awkwardly. "True" seals lack external ears, have shorter flippers, and move more like caterpillars when on land.

Able to see and hear underwater nearly as well as they can on land, seals and sea lions can stay submerged for long periods of time and descend to great depths. Elephant seals have reached depths of more than five thousand feet and remained underwater for ninety minutes. Because seals and sea lions dive so far below the surface of the sea,

Sea lions are highly social, and will often spend time together on beaches, rocks, and docks, sometimes snuggled up against one another.

they have been symbolically linked to the deeper parts of the mind, such as the unconscious. Furthermore, their life in the sea connects them to other symbolic aspects of water, including emotions, intuition, receptivity, and the imagination. Their amphibious lives inspired tales of shape-shifting and transformation beyond that of the selkie. Myths about the Norse trickster god, Loki, describe him as sometimes turning into a seal, and in Inuit legend, the goddess Sedna often appears with a seal's lower body.

Seals and sea lions are known for their impressive intelligence and irrepressible playfulness. They can be trained to respond to vocal and sign commands, can learn a variety of tricks, and have been used as performing animals in circuses and aquariums. Scientists recently learned that harbor seals navigate by the stars, and that sea lions are capable of abstract problem solving and making logical deductions. When it

Although we do not know as much about leopard seals—native to Antarctica—as we do about other seals, we do know that they, too, exhibit curiosity toward humans.

comes to having fun, these pinnipeds are up there with dolphins. They play games of chase and keep away, ride the surf, perform acrobatic stunts, and use other animals—such as starfish and pufferfish—as toys. Their behavior, both in the wild and in captivity, has made them emblems of playfulness.

INSIGHTS FROM THE SEAL AND SEA LION

Seals and sea lions have mastered the element of water, and in doing so call our attention to how we move through the more fluid territories of our psyches—our emotions, dreams, intuitions, imaginations, and unconscious. Do we spend enough time exploring and giving expression to these realms? Conversely, are we spending too much time in the watery aspects of self and need to find our footing? These amphibious creatures remind us to strive for balance between being solidly grounded on the shore (conscious, rational mind) and allowing ourselves to float in the amorphous fluidity of the sea (unconscious, emotional mind).

The seal's movement between the dark depths of the ocean and the sunlit surface also serves as a metaphor for bringing hidden things to light. Just as a seal needs to come up for air, unconscious content sometimes needs to break into the conscious mind. By giving this deeper material a passageway—through activities such as creative pursuits or meditation—we can tap into the richness of our inner worlds and achieve greater psychological integration.

Because seals are as associated with selkies as coyotes are with tricksters, their mythical role as shape-shifting creatures offers us additional insights. The most common version of the selkie story, told at the start of this chapter, tells us that without her magical skin, a selkie cannot transform back into her seal self. Separated from her pelt, she is out of her element, both literally, because she cannot return to the water, and figuratively, because she has lost a part of her true nature. Lacking her seal skin, she is forced to live in a world not her own, and under circumstances not of her choosing. It is only when she recovers her skin that she can transform back into her true self.

This story is a powerful metaphor for those times when we feel separated from deeply important aspects of ourselves, such our aspirations, faith, or sense of potential. Sometimes these aspects—our "seal skins"—are taken from us by circumstances beyond our control. Other times, we set them aside ourselves, intending to return to them, and then never do. Whatever the reason for our separation, when we are without access to these parts of ourselves, we are not whole. In this way, the seal glides through the depths of our psyches and surfaces to remind us to guard our skins, for—as long as we still have them—we can slip back into ourselves and remember who we are and where we belong.

The movement of seals and sea lions between the dark depths of the ocean and the sunlit surface of the sea can serve as a metaphor for bringing hidden things to light.

Ambition • Efficiency • Instinct • Motion • Predation
Restlessness • Stealth • Survival • Voracity

SHARK

A classic shark joke asks, "Why don't sharks eat lawyers?" and then delivers the punchline: "Professional courtesy." Most people assume that the word "shark" was first used for the fish and then as a derogatory term for ruthless attorneys. However, up until the sixteenth century, "shark"—which originated from the German *Schurke*, meaning "villainous scoundrel"—was used to describe "a dishonest person who preys on others," and was only later applied to the fish, presumably due to their predatory behavior.

So, first a shark was a scoundrel, then it was a fish, and when it became both, the two meanings cross-pollinated, resulting in the fish becoming more villainous and the villain becoming more shark-like. Today, the word shark is used to describe various questionable or potentially dishonest human behaviors, such as those represented by the idioms loan shark, card shark, and swimming with sharks. When applied to the aquatic animal, shark refers to a group of fish popularly perceived as fearsome, mindless eating machines. Scientifically, however, sharks are far more complex and intriguing, though some of their amazing adaptations have clearly contributed to their mythic status.

Found in oceans all over the planet, sharks range in size from the bioluminescent dwarf lantern shark, which can fit in a human hand, to the whale shark—the world's biggest fish, which grows to nearly sixty

Sharks can become so focused on feeding that they seem to lose their minds, often thrashing against—and sometimes even chomping on—one another.

feet in length and weighs as much as fifteen tons. Although there are five hundred species of sharks, all share certain anatomical characteristics: they have hydrodynamically sleek body shapes; flexible, lightweight skeletal structures made of cartilage instead of bone; and skin covered in friction-reducing dermal denticles instead of scales. These attributes enable them to swim fast and turn quickly, which make them efficient hunters. Some species of sharks need to swim constantly in order to keep oxygen-rich water flowing over their gills. If they stop swimming, they can't breathe and drown. The shark's perpetual motion links him to the ideas of relentless pursuit, ambition, and restlessness.

Perhaps more than any other feature, it's the shark's impressive set of razor-sharp teeth that has given him a reputation as a ferocious predator. His teeth are positioned in multiple (five to seven) rows inside his jaw. When a tooth is knocked out—which happens often with a biting force of up to eight thousand pounds per square inch—the corresponding tooth in the reserve row moves forward to replace it. It is no wonder that sharks are so often reduced to a cliché represented solely by their fearsome jaws.

The shark's senses are no less effective than his speed, agility, and bite. He can smell as little as one drop of blood in a million drops of water and from at least a quarter mile away. He can detect low-frequency sounds, such as those emitted by injured or sick prey. With eyes positioned on the sides of his head, he has a nearly 360-degree field of vision and sees exceptionally well in low light. His eyes also are highly sensitive to the fluctuations of light and shadow that occur when prey moves through the water.

If all their other predatory advantages weren't enough, sharks have two extra senses. They have an enhanced sense of touch known as a lateral line—comprised of small pressure-sensitive pores that run from snout to tail—that enables them to sense pressure changes in the water, such as those created by prey or predators. They also have electroreceptor organs located on their snouts and heads that detect the electromagnetic fields generated by muscle contractions in nearby fish. Sharks use their electroreceptors to find prey in the dark or even when hidden under the sand.

Sharks often appear out of nowhere and then disappear just as quickly, connecting them to the concepts of stealth, elusiveness, and the element of surprise. When they have honed in on prey, a feeding frenzy often follows. They can become so focused on feeding that they seem to lose their minds, often thrashing against—and sometimes even chomping on—one another. This behavior made sharks symbols of voracity and competition—and, by extension, unchecked ambition. Some sharks don't even wait until they are born before they start competing. In several species, the first embryo to hatch eats the unhatched eggs. The sand tiger shark takes a different approach: females have two uteri and the largest embryo in each uterus waits until his siblings have hatched and then cannibalizes all of them, leaving the female to give birth only to the strongest pup in each womb.

The shark's anatomy is so perfectly designed for sensing, hunting, and devouring prey that he epitomizes efficient predation. However, as scientists learn more about sharks, a more complex image of these fascinating fish is emerging. For instance, many sharks cooperate, socialize, and form bonds with one another. They demonstrate curiosity and intelligence, have long memories, and are capable of problem solving. Scientists have even trained sharks to recognize shapes and colors and to roll over for belly rubs.

Cultures that live in close proximity to sharks have regarded them very differently from those in the West. Many Polynesian cultures consider the shark sacred and endow him with magical powers. Throughout parts of Polynesia, when people look up at the Milky Way, they see a shark known as the "Long-Blue-Cloud-Eater." In Hawaiian mythology, Kamohoalii was a beneficent shark god who, in exchange for a drink of kava (a mildly intoxicating beverage), would guide lost sailors back to their ports. Hawaiians also believed that if you offered your deceased family member's corpse to a shark, that person would transform into a shark who would then act as a guardian spirit, protecting the family and helping them catch fish.

Various species of sharks have been swimming in Earth's seas for more than four hundred and fifty million years. Although sharks have evolved during this time, their fundamental anatomy has remained relatively consistent, linking them to the concepts of endurance and survival. But their reputation as survivors could change soon, as humans are killing roughly one hundred million sharks globally each year. As for the sharks, they kill fewer than ten people a year. So who's the true *Schurke*?

Sharks often appear out of nowhere and then disappear just as quickly, connecting them to the concepts of stealth, elusiveness, and the element of surprise.

INSIGHTS FROM THE SHARK

Over millions of years, the shark has evolved a perfect set of adaptations to succeed in his ecosystem. He is the steely, efficient survivor—the no-nonsense king of the sea—who connects us with our own ability to persevere and circle to the top of our domains. He is streamlined, which reminds us that when we assume a form or attitude that presents little resistance, it is easier to move through life without turbulence. Being flexible, he shows us the value of being able to instantly change our direction and focus. His ability to replace teeth—symbolic of power and vitality—invites us to consider how we, too, might cultivate inner reserves of strength and stamina.

Those species of shark who need to swim constantly to keep from drowning offer us a metaphor for our own need to keep moving forward in order to thrive. They connect us with our drive and ambition, reminding us that if we want to achieve our goals, it's rarely wise to hesitate or sit on the sidelines. Using their acute senses, sharks are able to detect opportunities, which reminds us to hone our own senses and, as we read the room looking for opportunities and watching for obstacles, trust our intuitions and instincts.

Conversely, sharks are sometimes so carried away by their drive to feed that they stop at nothing, even if it means inadvertently attacking their own kind in the process. Some sharks have even beached themselves during feeding frenzies. Sharks, therefore, caution us not to get too caught up in our own ambition and competitive impulses lest we lose sight of the bigger picture, endanger ourselves, and hurt others.

For nearly five hundred years, Western cultures have been attributing some of humanity's most detestable traits to sharks, despite the fact that most of the negative qualities we attribute to sharks only apply to humans, not to the fish. So instead of perceiving this majestic, ancient fish as a shark—or *Schurke*—perhaps we should envision him as the Long-Blue-Cloud-Eater—a fascinating fish that we just might get to know better if we can see past our projections.

Perhaps more than any other feature, it's the shark's impressive set of razor-sharp teeth that has given him a reputation as a ferocious predator.

**Community • Fertility • Gentleness • Innocence
Purity • Sacrifice • Vulnerability**

SHEEP

THE SHEEP IN NATURE AND CULTURE

In 350 BCE, the Greek philosopher Aristotle opined that the sheep is "naturally dull and stupid. Of all quadrupeds it is the most foolish." Two thousand years later, George Washington, the Commander in Chief of the Continental Army (who later became the first President of the United States), expressed a similar view of sheep when he told his officers that if freedom of speech were taken away, then "dumb and silent we may be led, like sheep, to the slaughter." More recently, in 2017, Merriam-Webster announced the addition of the word "sheeple"—defined as people who are docile, compliant, or easily influenced—to its dictionary.

Sheep were domesticated roughly ten thousand years ago, most likely bred from the mouflon, a wild sheep native to Asia and Europe. One of the most farmed animals on Earth, sheep have provided humankind with wool, hides, meat, and milk. We don't know at what point during our relationship with sheep we became so convinced of their lack of intelligence that they came to represent dim-witted passivity, but there's no doubt that the idea has endured through the ages. Like their close relatives, goats, sheep are grazing herbivores with horns and cloven hooves. They differ from goats by having tails that hang down (the goat's usually points up); thicker, curved horns on the sides of their heads (goat horns are thinner, straighter, and on top of their heads); and no beards.

In Western cultures—which generally equate individualism and initiative with intelligence—the sheep's self-effacing assimilation into the flock resulted in the perception that they were dull, helpless animals.

Another difference between sheep and goats is the way they forage. Staying with their flocks, sheep graze on plants close to the earth, and sometimes become so attached to their pastures—a phenomenon called hefting—that they can be kept without fences. Goats, in contrast, tend to wander off on their own to browse on taller plants. The difference in foraging habits between sheep and goats contributed to the sheep's symbolism. Compared to goats, who are more independent and adventurous, sheep—with their strong flocking instincts—were seen as conformists. They were perceived as meek, mild, and unassuming. They didn't stand out in a crowd like goats, who often call attention to themselves with their bold antics. Instead, sheep seemed to submissively surrender their will and relinquish the ability to think and act for themselves. In Western cultures—which generally equate individualism and initiative with intelligence—the sheep's self-effacing assimilation into the flock led to the perception that they were dull, helpless, and vulnerable.

Even the sheep's compliance—seen as a sign of intelligence in dogs and horses—was interpreted as evidence of the sheep's innate gullibility, except when viewed through the lens of Judaism and Christianity. In these religions, the sheep's submissiveness was equated with trust in and obedience to God. The Hebrew Bible often describes God as a shepherd and refers to his followers as sheep. The Twenty-third Psalm, which begins "The Lord is my shepherd," is one of the most familiar examples of this metaphor. Later, in Christianity, the expression continued to be used, this time to describe Christ as the "Good Shepherd" who leads his flock of followers to God. Within Judaism, Christianity, and Islam, lambs were often ritually slaughtered to appease God, which is why a "sacrificial lamb" is a metaphor for a person or animal sacrificed for the common good.

Outside of religion, the word shepherding—literally defined as guarding sheep and preventing them from straying—is often used to describe the act of leading, guiding, and protecting a group of people. As a symbol, the shepherd is always tied to the idea of the flock—a collective of individuals who are in some way vulnerable or aimless and require guidance. The flock metaphor also inspired the ideas of the lost sheep— a person without direction due to separation from the flock and the loss of guardianship of the shepherd, and the black sheep—a person shunned because of deviation from the accepted standards of her group.

Sheep have additional associations derived from gender and age. Rams are known to behave passionately and aggressively during breeding season, often challenging one another for hierarchy and breeding opportunities, which made them symbols of impetuousness, determination, and virility. No doubt this is why Khnum, the ancient Egyptian god of fertility, was represented with a ram's head. When the word ram is used as a verb, it describes the action of forcefully charging forward and striking something. In contrast, ewes are usually gentler and more docile. They are tender and devoted mothers who form deep and lasting bonds with their lambs. As for the lambs: their sweet, gentle demeanor has made them an enduring symbol of purity and innocence. Because they are born in the spring when life is returning to the world, they also are associated with fertility, renewal, and hope.

Although sheep have long been perceived as mindless, they aren't. First, the very flocking behavior that gave them the reputation for being empty-headed is, in reality, an effective survival mechanism that protects them from predators. Sheep have eyes positioned on the sides of their heads, which, along with their horizontal pupils, give them a very wide visual field—up to 320 degrees. Although an individual sheep on her own has a broad view of the surrounding environment, an entire flock becomes an incredible surveillance system. With so many sheep watching for predators and alerting one another to approaching threats, the flock greatly increases its chance of survival.

The metaphor of the black sheep—a person shunned because of deviation from the accepted standards of her group—arises out of flock mentality.

Perhaps just as importantly, flocking behavior not only keeps sheep safe, but has contributed to their impressive emotional and social intelligence. In fact, research scientists have discovered that sheep demonstrate certain kinds of intelligence on par with that of primates. They have impressive memories and can recognize and remember at least fifty individual sheep faces for more than two years, and can differentiate the emotional states of flock mates through facial expressions. They also forge long-term friendships, demonstrate loyalty, defend one another in disputes, and exhibit stress when seeing friends taken to slaughter. For sheep, life is all about relationships—who you can trust and rely on to alert you about predators, and who you can peacefully spend time with.

INSIGHTS FROM THE SHEEP

Lambs call our attention to the aura of potentiality and hope that imbues new beginnings. They evoke memories of innocence and vulnerability, and especially the comfort of being shepherded by loving parents, teachers, and others who looked out for us when we were children. The lamb, however, also offers a melancholy reminder that as we grow into adulthood—and continue through the stages of life— we have to periodically relinquish an "innocence of heart" in order to gain the knowledge and maturity we need to forge a life as individuals, trusting ourselves and our choices.

Sheep teach us that there is comfort and safety in belonging to a community—that in so many aspects of life, "united we stand, divided we fall." The flock is a reflection of the value of strong bonds, loyalty, and friendship. And a hefted flock—one that is so imprinted on a particular place that the sheep do not need to be fenced—invites us to embrace the pleasures of feeling rooted in and connected not only to others but also to a place, especially one we call "home."

Rams are known to behave passionately and aggressively during breeding season, often challenging one another for hierarchy and breeding opportunities, which made them symbols of impetuousness, determination, and virility.

Conversely, we need to resist losing ourselves in the flock, forgetting to think and act on our own for fear we might upset the status quo. In his allegorical novel *Animal Farm*, George Orwell cast the sheep in the role of the duped citizens of a totalitarian system, reminding us of the dangers of conformity and "group think." When we over-identify with any single group—as a "dyed in the wool" member of a political party or religious group—we can become inflexible and dogmatic. Stuck in a group mindset, we're more inclined to judge those who step out of line or simply don't fit in, calling them "black sheep" and ostracizing them from the flock.

So, while there is safety and comfort in numbers, there is also the risk of losing the self. If we trust blindly, we can too easily be taken advantage of by the "wolf in sheep's clothing." The dichotomies associated with sheep—from lamblike innocence and group togetherness to loss of individuality and fear of being cast out—asks us to consider all the ways we benefit by balancing our need for autonomy with our desire to be a valued member of a family and community.

Sheep teach us that there is comfort and safety in belonging to a community—that in so many aspects of life, "united we stand, divided we fall."

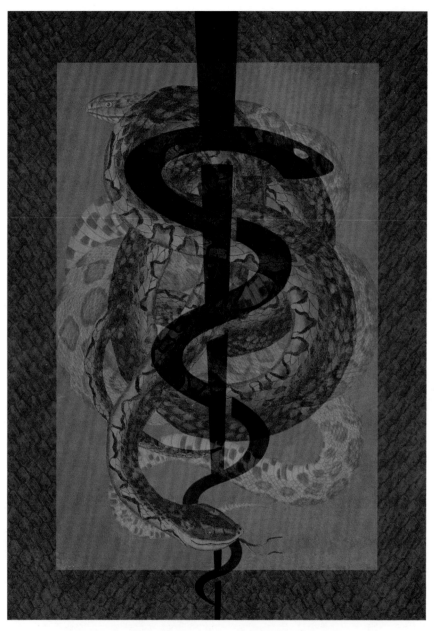

Creativity • Deception • Destruction • Duality • Evil
Fear • Fertility • Healing • Immortality • Life Force
Rebirth • Sexuality • Water • Wisdom

SNAKE

THE SNAKE IN NATURE AND CULTURE

In the beginning was the snake. She was male and female, life and death, chaos and order, good and evil, creator and destroyer. In her serpentine S-shape, she was the forward-flowing river of time. Looped into a circle, she was Ouroboros—the serpent who eats her own tail—signifying the endless cycle of time flowing back into itself. By embodying all these dualities, the snake was both energy and matter, potential and manifestation.

Found nearly everywhere on Earth, the snake is the most ancient and widespread animal symbol, dating back to the Stone Age. The oldest evidence of the snake's presence in the human imagination might also be the earliest evidence of religion. In 2006, in a remote hill cave in Botswana's Kalahari Desert, archaeologists discovered a twenty-foot-long stone carving of a python. They believe that the python, and more than a thousand other artifacts discovered at the site, were carved roughly seventy thousand years ago by ancestors of Botswana's San people, who are considered to be one of the most ancient of human cultures. Even today, the python is an important animal to the San, whose creation story describes humankind as descending from this snake. Archeologists speculate that firelight reflecting off scale-like facets carved along the python's body could have created the illusion of movement, giving the stone serpent a supernatural aura.

When ready to take action, snakes are as bold as a flash of lightning, striking without hesitation when in harm's way or when an opportunity is in reach.

Archeological and historic records suggest that the snake has always evoked both reverence and fear in the human heart, although fear has largely eclipsed reverence in modern humans. Some scientists believe that a fear of snakes might even be hardwired in our DNA. The theory is based on the premise that primates who learned to avoid snakes were less likely to be killed by them, and therefore lived long enough to breed. Over time, through natural selection, a fear of snakes became an inherited biological trait in all primates, including humans.

Whether hardwired or learned, humankind's fascination with and fear of snakes has found expression in mythology, religion, folklore, philosophy, psychology, and other aspects of culture. First appearing in the Paleolithic period, the human obsession with snakes slithers through time, appearing nearly everywhere humans live. Early peoples associated snakes with water—necessary to all life—so snakes became linked to fertility and the primitive earth goddesses of prehistory. The snake also is found in numerous creation stories, such as those of the Australian Aboriginal people, which describe a Rainbow Serpent who gave birth to all the animals.

The snake also entwined herself with the more nuanced goddesses of early cultures, such as Isis, the ancient Egyptian goddess of healing and magic, who sometimes took the form of the cobra goddess, Isis-Renenutet. In this manifestation she had a cobra's head, and could vanquish enemies with her gaze. Similarly, in Greek mythology, the monstrous Medusa—a creature with snakes for hair—could turn men into stone simply by staring at them. This mythical ability was likely inspired by the biological cobra, who was believed to hypnotize prey with her eyes.

In the Judeo-Christian story of the Garden of Eden, a talking serpent—a trickster figure later associated with Satan—tempted Eve into eating the forbidden fruit from the tree of knowledge. As punishment, Adam and Eve were banished from paradise, childbirth became a hardship for women, men had to work the soil, and the serpent was cursed to crawl on his belly and eat dust for all eternity. This story reflects the serpent's association with evil within the Judeo-Christian tradition, as well as her link to sexuality, which arises from her association with fertility and the life force, her phallic form, and her unusual breeding behavior. Some species mate in a "breeding ball" that consists of one female and numerous males who stay grouped together for up to a month. In Hinduism, Kundalini, which means snake, is the term for sacred sexual energy, which is visualized as a serpent coiled at the base of the spine.

Snakes smell through their flickering, forked tongues, sense vibrations through their bellies, hear through their jaws, and some even have

The snake has been associated with healing since ancient times, which is why Asclepius, the Greek god of medicine, carried a staff with a serpent coiled around it.

infrared receptors, which enable them to detect the radiating heat of warm-blooded prey. These unusual perceptual abilities link them to the idea of secret wisdom, as does their descent into and emergence from openings in the earth—the "underworld." Because people believed that snakes could travel between worlds, snakes were presumed to have access to unearthly knowledge.

Snakes often appear unexpectedly and slip along surreptitiously, which is why we associate them with stealth, secrecy, deceit, and the idea of sudden danger. This gave rise to idioms such as "slippery as a snake," "snake in the grass," and "speaking with a forked tongue," in which the snake designates deceit and cunning. Although snakes are known for their fluid riverine slither, they actually move in a wide variety of ways—sliding, inching, leaping, sidewinding, swimming, and gliding. They also strike, often suddenly and with great speed, which links them to the zig-zag path of lightning. The word snake is not just a noun—it's a verb, too, which reflects the animal's association with energy in motion.

When snakes strike, it's usually to sink their teeth into prey or predators. Venomous snakes inject venom through their fangs or spit it at prey, whereas constricting species use their muscles to coil around and suffocate their prey before swallowing it whole. To accomplish this feat, their jaws unhinge, enabling them to wrap their mouths around prey. They then use their teeth and muscles to pull the food toward their stomachs, where it digests, usually for several days. The eerie, uncanny way that snakes kill and consume their prey taps into our deepest fears of being devoured, which is why snakes came to symbolize destructive forces.

Even though only two hundred of the three thousand snake species are venomous enough to seriously harm or kill a human, their potential to do so inspired ancient cultures to see them as sinister and supernatural. But the snake's venom, like the snake herself, embodies duality: it harms *and* heals. Antivenom—a drug given to stop the toxic effects of snake venom—is made from the very same venom that causes the injuries and fatalities. However, long before people knew how to make antivenom, the snake was already associated with healing, through her alleged ability to travel to the underworld and her periodic skin shedding, both of which represent renewal, regeneration, rebirth, and immortality.

In Greek mythology, Asclepius, the god of medicine, carried a staff with a serpent coiled around it. One story tells that a serpent whispered the secrets of the healing arts to him. Another describes Asclepius observing one snake feeding herbs to a second, dead snake and reviving him in the process. Asclepius took note and later used the same herbs to revive a dead man. Asclepius's serpent-entwined rod became an emblem of healing that is used around the world today. A similar

The snake's unusual perceptual abilities link her to the idea of secret wisdom, as does her descent into and emergence from openings in the earth—once considered access points to the "underworld."

symbol of medicine, the caduceus—two snakes intertwined around a rod with wings at the top—originated from the staff of the Greek god Hermes, messenger of the gods.

INSIGHTS FROM THE SNAKE

The snake's association with dualistic ideas offers us a helpful framework for dealing with the discord we feel when pulled in two directions. When we find ourselves needing to make a difficult choice, we need to remember that whichever decision we make, chances are the consequences will not be either/or—there will be regrets *and* fulfillments. If we can make peace with the opposing energies, then either decision has greater potential to make us happy. In this way, the snake shows us that accepting duality can help us to feel less fractured and more unified as we forge a path forward.

Often compared to a flowing river, the snake is the epitome of flexibility, gliding with ease over most terrains. As such, she invites us to think about whether we move through the world with either fluidity or resistance. When ready to take action, snakes are as bold as a flash of lightning, striking without hesitation when in harm's way or when an opportunity is in reach. The snake does not waste time. Surreptitious and slippery, the snake also cautions us to watch for those speaking with forked tongues or otherwise behaving in duplicitous ways. Furthermore, as a creature who swallows her prey whole, the snake warns us to be careful about what stories we "swallow" without question or scrutiny, lest we find ourselves with a belly full of regret.

Finally, the way snakes shed their skin offers us an age-old metaphor for sloughing off perspectives and patterns that no longer serve us. Sometimes our "sheddings" are elective—we decide to give up a job, relationship, habit, or other aspect of self. Other times, circumstances—such as an illness or unexpected loss—force us to change. Either way, the snake is a reminder that each of us, like the serpent herself, is both a noun and a verb, forever coiled in a circle of transformation and emergence.

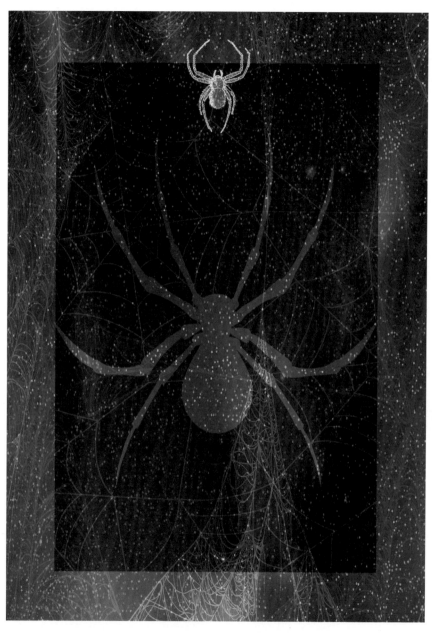

**Creativity • Deception • Destruction • Fate • Fear
Patience • Predation • Time • Wisdom**

SPIDER

THE SPIDER IN NATURE AND CULTURE

According to prevailing cosmological thinking, billions of years ago there was only a single, tiny point that contained all matter, energy, space, and time. Then it spontaneously expanded, creating the universe. As it stretched outward—in interconnected filaments separated by empty space—radiant galaxies of stars and planets formed at the intersections of the celestial threads. Scientists call this structure the cosmic web, and not only does it bear a resemblance to spider webs, but it also reinforces the spider's age-old association with creativity.

Long before astrophysicists discovered the web-like structure of the universe, cultures around the world associated the spider—a tiny creature who is the point of origin of her own web—with the act of creation. In Hinduism, the spider and her web are connected to the metaphysical concept of the Bindu, the origin point from which the universe and consciousness originates. Within the traditions of the Hopi, Pueblo, and Navajo peoples of the American Southwest, the spider is personified as Spider Grandmother (sometimes called Spider Woman), who created the world by thinking it into existence while weaving her webs. In West Africa and the Caribbean, we find Anansi, a creator god and trickster in the form of a spider, who created the sun, moon, and stars, and even the material from which humans were fashioned.

In ancient Egypt, the spider was associated with Neith, the goddess of weaving and creation, who used her loom to re-weave the universe into existence every day.

With more than forty-eight thousand species distributed over every continent except Antarctica, the world is filled with spiders. In fact, scientists determined that in natural settings, we are never more than three feet away from a spider—a fact that probably doesn't sit well with those suffering from arachnophobia. Given the spider's otherworldly features, it's easy to understand why they evoke fear in so many people. Members of the class of animals known as arachnids, spiders are characterized by having eight legs, two body segments, fangs, multiple eyes (usually eight), and spinneret organs that produce silk. Spiders use silk for trapping, wrapping, transportation, cocoons, nests, and to protect offspring. With very few exceptions, spiders are carnivorous, and either catch prey in their webs, ambush them, or stalk and pounce on them.

Nearly all spiders have venom glands and fangs. Spiders don't chew their prey, but instead inject it with digestive fluids. After the prey is liquified, the spider sucks out its remains. The females of some species, such as the Black Widow, occasionally consume males after mating, which links the spider to the idea of the treacherous femme fatale. With such an eerie and terrifying style of predation, it's no surprise that spiders elicit feelings of revulsion and serve as inspiration for monsters of science fiction and mythology. In Japanese folklore, for example, a race of gigantic spiders known as *tsuchigumo* prey on humans, reflecting our archetypal fear of the spider's destructive energies—the counterpoint to her creativity.

Because spiders spin silk and weave webs, they have long been connected to the craft of weaving, which symbolizes creation and fate. In ancient Egypt, the spider was associated with Neith, the goddess of weaving and creation, who used her loom to re-weave the universe into existence every day. In Greek mythology, Arachne, a gifted weaver—from whom we get the spider's scientific name—tempted fate by challenging Athena, the goddess of wisdom, to a weaving contest. The goddess prided herself on her weaving skills and became infuriated by Arachne's impertinence when the mortal outperformed her. Athena tortured Arachne to the point of such despair that she hung herself. The goddess then took pity on Arachne and brought her back to life as a spider so that she could weave for all eternity. This myth reflects another quality that humans have projected onto the creature who sits at the center of her own creations: arrogance and conceit.

The spider's fate, like Arachne's, is interwoven with her threads. Not all spiders make webs, but they all produce silk, a material stronger than any known fiber—natural or synthetic. Spider silk begins as a liquid in the spider's body and is then spun into fiber by spinnerets—the spider's silk-spinning organs. All spiders produce a variety of silk, some of which is sticky. Spiders who make webs do so in a variety of patterns and shapes, from the archetypal wheel-shaped orb web that traps flying insects to webs shaped like tangles, trampolines, funnels, tubes, ladders, and other geometric forms.

Spiders call our attention to the risks of getting stuck in the proverbial tangled webs we weave, especially when we—consciously or unconsciously—set traps for one another to prove a point or otherwise control a conversation.

As intriguing as their webs are, spiders build them primarily to catch prey, which is why they are also associated with deceit. Mary Howitt's famous poem "The Spider and the Fly" captures this aspect of spiders with the memorable opening line, "'Will you walk into my parlor?' said the spider to the fly." This line became a metaphor for a false offer of kindness that is, in fact, a trap, and reinforced the perception of spiders as masters of ensnarement.

Howitt's characterization of spiders is fair, as they have a repertoire of especially sinister ways to capture prey. Some species spin webs and wait for prey to get stuck in them, relying on sensors in their legs to alert them to the presence of their captives. Others set trip lines and then wait until vibrations in the line indicate it's time to pounce. The Trap Door spider digs a burrow and spins a circular door that she attaches to the opening. When prey walks over the door, she bursts through the silk and ambushes it. The Ogre-faced spider weaves a small rectangular web and then casts it like a net onto her unsuspecting victim.

One of the most ingenious feats of spider engineering is performed by the Slingshot spider, who weaves a conical web with a single strand attached to the center. She reels in this line to create tension and, when an insect flies by, she releases the strand, shooting herself and her web at the target. With her many ways of luring prey into her parlor, the spider came to symbolize wisdom, deceit, and cunning. And like the spider in Howitt's poem, who doesn't give up until the fly gives in (it takes six stanzas to describe all the schemes the spider employs in her efforts to seduce the fly), she also represents patience and persistence.

Another thing that spiders do with their silk is build shelters, with one of the most interesting belonging to the Diving Bell spider, who spins a bell-shaped web underwater, anchored on plants. She then collects tiny air bubbles from the surface of the water and combines them into one giant air bubble that floats at the top of her diving bell, providing her with oxygen while she waits for small fish, insects, and larvae

Scientists have determined that, in natural settings, we are never more than three feet away from a spider—a fact that probably doesn't sit well with those suffering from arachnophobia.

to cross her path. Spiders use silk for transportation, too. Jumping spiders create a silk safety line when leaping long distances. Several species "balloon" by releasing strands of silk and letting the wind catch and carry them. Some stretch a silk thread between two points so they can travel across it like a bridge. This connecting aspect of the spider's threads is featured in myths in which the spider is depicted as spinning silk to connect heaven to earth, and then playing the role of mediator between the gods and humanity.

INSIGHTS FROM THE SPIDER

Whether creating webs or setting trip lines, spiders are masters of deception and entrapment, and so call our attention to the risks of getting stuck in the proverbial tangled webs we—and others—weave. When communicating, for example, threads of conversation can get sticky, especially when we—consciously or unconsciously—set traps for one another to prove a point or otherwise control the exchange. The spider warns us to watch where we step and do our best to keep a path clear for others. No one likes walking into a spider's web—literal or metaphorical.

Many species of spiders begin weaving their webs by securing lines to plants, trees, or other objects in the environment, reminding us that we, too, need to anchor ourselves within the larger contexts of community and world. A spider sitting smugly in the center of her creation—her web—can evoke an image of someone claiming the center of attention. In this way, the spider reminds us that it's fine to feel pride in one's work, but not at the cost of getting tangled in conceit or stuck in solipsistic introversion. More often, the image of a web with a spider at its center evokes the idea of each of us as the creators of our own lives. Despite sometimes feeling like we are trapped in a web not of our own making, we are never solely the victims of circumstance. The threads we spin and the patterns we weave shape our journeys, so the spider advises us to look carefully at how we create—or at least contribute to—the fabrics of our lives.

In E. B. White's classic children's book *Charlotte's Web*, a barn spider named Charlotte tries to save a livestock pig named Wilbur by writing positive messages about him in her web, hoping to persuade the farmer to spare his life. White, however, does not soften Charlotte's true spider nature. Instead, Charlotte openly admits to using her web to trap insects whose blood she drinks for dinner, shrugging it off as she tells a horrified Wilbur, "A spider has to pick up a living somehow or other." And yet, Charlotte uses her web—the same one she traps insects with—to help Wilbur. Charlotte shows us that spiders—and humans—embody both creative and destructive energies. We all belong to the great web of life and we all devour something, whether lettuce or flies. This is the price of being alive. However, when it comes to our creativity, we all have our silk, but what we choose to spin with it—a mandala that glistens in the morning light or a sticky web to entrap another—is up to us.

Mary Howitt's famous poem "The Spider and the Fly" memorialized the perception of spiders as masters of ensnarement with the opening line, "'Will you walk into my parlor?' said the spider to the fly."

**Beauty • Elegance • Fidelity • Inspiration • Light • Love
Poetry • Purity • Spirit • Transformation**

SWAN

THE SWAN IN NATURE AND CULTURE

A legend from the Ainu, the indigenous people who live on Japan's Hokkaido island, tells us that the swan originally lived in paradise as an angel. A long time ago, when the Ainu were on the verge of extinction, the swan took the form of a woman who came to Earth to save their people. Even today, many of the residents of Hokkaido still regard swans as visitors from heaven, which is why they call the migrating swans who arrive as the first snow falls the "angels of winter."

With their spectacular white plumage and expansive wingspan (up to ten feet), swans have a mythical quality and have been seen as spirit guides, angels, and companions of the gods since ancient times. Archeologists believe that the perception of swans as spirit guides dates back to Paleolithic times, a theory supported by the discovery of twenty-two thousand year old swan talismans in grave sites. The swan's ethereal appearance, his ability to move between three elemental worlds—air, earth, and water—and his mysterious migratory arrivals and departures might have inspired the belief that swans can carry spirits to the afterlife.

Often contrasted with the goose—who is somewhat more mundane in appearance—the swan, with his snowy white feathers and gracefully curved neck, has long represented purity, refinement, and elegance.

It can take a year or longer for cygnets to develop their magnificent adult plumage, and when they do, it's easy to see how their stunning transformation inspired Hans Christian Andersen's fairy tale "The Ugly Duckling."

The high regard people had for swans can be found in the proverb "to turn geese into swans," in which the swan stands for the sublime, while the goose represents the ordinary. Due to their exquisite appearance and lofty associations, swans have been associated with the arts since antiquity. In Greek mythology, swans were sacred not only to Apollo, the handsome god of poetry, music, and art, but also to the Muses, the goddesses of inspiration. Over time, the word swan came to be used as an epithet for skilled poets, as well as anyone possessing extraordinary talent or beauty.

The beauty of swans is reflected even in their mating rituals. Known to be one of the most monogamous creatures on the planet, paired swans engage in a slow courtship dance while holding each other's gaze. Facing each other, they spread their wings and ruffle their feathers, often bowing head to head, creating a perfect heart shape in the space between them. Just before consummating their bond, they intertwine their necks in what strikes many as a romantic gesture of tender affection. Their courtship dance, along with their tendency to mate for life, made swans symbols of love and devotion in many cultures.

Together, the bonded pairs raise many clutches of young, called cygnets, who emerge from their eggs covered in a light gray, fluffy down. It can take a year or longer for cygnets to develop their magnificent adult plumage, and when they do, it's easy to see how their stunning transformation inspired Hans Christian Andersen's fairy tale "The Ugly Duckling." When young, cygnets are vulnerable and need protection. Both parents are known to boldly attack intruders when caring for their cygnets. Males, in particular, will fiercely guard their nests, protecting their young from predators and their mates from other males. If necessary, males will rear up with flapping wings, grasp with their beaks, and kick with their legs to ward off a threat. The swan's surprising strength and courage only added to his reputation as a noble and virtuous creature.

Swans are also associated with shape-shifting and transformation. There are many myths and folktales about swans who change into humans, including the well-known Greek tale of Zeus, the god of all gods, turning himself into a swan of such radiance that he was able to seduce Leda, the queen of Sparta. But the most widespread swan transformation story—found throughout the world—is that of the swan maiden. Although details vary, the essence of the story is essentially the same. A hunter stumbles upon a group of beautiful maidens bathing in a lake, notices their feathered robes on the shore, and steals one. When the women return to shore to put their cloaks on and turn back into swans, one finds her robe missing. The hunter then coerces her into marriage and she bears his children, but at some point she recovers her robe, returning to her true nature, and flies away.

Another widespread folk belief—that dying swans produce a beautiful song-like sound just before they expire—connects the swan

Swans present a romantic image during their mating rituals, in which paired swans engage in a slow courtship dance while holding each other's gaze.

to life's final transition: death. The expression "swan song," which arose from this legend, became an idiom referring to a final performance or accomplishment that occurs before retirement, death, or other ending. Although the literal swan song has been largely dismissed as folklore, researchers have found that Whooper swans sometimes produce a drawn-out series of notes as their lungs deflate upon death.

When we imagine a swan, our conjured image of this bird will no doubt be white. But there are actually black swans, which are found in Australia, Tasmania, and New Zealand. Because they are uncommon, the black swan has become a symbol of rarity as well as the inspiration for a philosophical concept about how rare and unexpected events can have startling, even extreme, impacts on our understanding of the world. The black swan serves as a reminder not to overvalue orthodoxy, as doing so limits creativity and innovation.

INSIGHTS FROM THE SWAN

The swan's allure arises in part from the way he embodies contradictory extremes, such as grace and power, and so invites us to embrace opposing and sometimes conflicting qualities in ourselves. When we allow contrasting qualities—such as passion and objectivity—to coexist in ourselves, we develop a healthy system of checks and balances. So, let your grace be powerful and your power be graceful.

When two swans face each other as they float together on the water, their heads bent and beaks just barely touching, they create a heart-shaped space between their two bodies. This powerful image—of two halves of a heart coming together as one—evokes the enduring idea that each of us seeks the missing half of ourselves in another in order to realize our wholeness. While we may or may not believe in the concept of the "soul mate," swans nonetheless inspire us to embrace the mystery of those relationships that expand—and sometimes even complete—our hearts.

The cygnet's transformation from a fuzzy, gray hatchling to a regal white swan—along with the fairy tale it inspired—reminds us that it can take time to figure out who we are and to which flock we belong. When the little cygnet becomes a swan and finally recognizes himself as one, he brings us back to those moments in life—not only when we were young, but throughout our journeys—when we recognize our own worth and place in the world. And when circumstances cause us to feel estranged from ourselves, the swan maiden tale offers us yet another hopeful message. Despite her enforced marriage to a mortal, the swan maiden never forgets who she really is. With patience, she finally frees herself, finds her feathered robe, reclaims her true nature, and soars into the sky.

**Confidence • Drive • Elusiveness • Majesty
Mastery • Power • Predation • Strength**

TIGER

THE TIGER IN NATURE AND CULTURE

"Tyger, Tyger, burning bright / In the forests of the night" are the now-famous words with which William Blake opens his poem, "The Tyger." The poem proceeds to ask who could have created a creature as fearsome as the tiger—an animal so frightening it must have been fashioned from fire. As the poem progresses, it concedes that it could only be God's hand that fashioned the fiery beast, even though we'll never know his reasons. The poem ultimately yields to the truth of the tiger: she is both terrifying and beautiful, dangerous and wondrous—like creation itself.

Blake's poem reflects the profound impression the tiger has made on the human imagination. We are awed by the tiger, and how could we not be? She is the largest cat in the world, growing to ten feet in length and weighing as much as eight hundred pounds. She is able to leap more than thirty feet in a single bound, run at speeds of forty miles an hour, climb trees, and swim with surprising ease. No wonder the tiger symbolizes power and strength.

The tiger steps silently through the world on padded paws that conceal claws up to four inches long. She finds her way in the dark with superior night vision, eyes flashing like fire. As for her hunting strategy, the tiger combines the qualities of patience, perfect timing, and

In Southeast Asia, people believe in a "tiger ancestor" who leads initiates into the jungle where they are symbolically killed and brought back to life as shamans with the ability to shape-shift into tigers.

remarkable swiftness to take down prey more than twice her size. She is calm, cool, and collected . . . until she is not. In Tibetan Buddhism, the tiger symbolizes unconditional confidence and disciplined awareness, both of which were undoubtedly inspired by the way tigers appear to be relaxed at all times, yet ready to strike at any time.

Wherever the tiger lives, whether in the snowy birch forests of Siberia or the mangrove forests of India, she is the apex predator, the master of her domain, and the indisputable queen of her ecosystem. This is why the stripes on the tiger's forehead inspired the Chinese character *Wang*—which translates as "king"—and why cultures throughout the tiger's "kingdom" saw her as the embodiment of mastery and majesty.

And then there's the tiger's arresting physical beauty, which is made all the more alluring by her ability to vanish in the blink of an eye. Surprisingly, the tiger's conspicuous coat—the same one that makes her stand out—is, in fact, a remarkable adaptation. It enables her to blend into foliage and merge into shifting patterns of shadow and light like a master illusionist. One minute the tiger is there, staring at you from a field of grass; the next she is gone.

Everything about the tiger—from her imposing physical presence and predatory prowess to her elusiveness—inspired those who share their world with the tiger to attribute supernatural powers to her. For example, the indigenous peoples of Siberia have long attributed magical powers to tigers, such as the ability to command other animals to make themselves easy prey for hunters. According to one folk belief, tigers also help out those who have lost their way in the forest. For example, hunters who wander off the trail and become disoriented

Surprisingly, the tiger's conspicuous coat—the same one that makes her stand out—is, in fact, a remarkable adaptation that enables her to blend into foliage and merge into shifting patterns of shadow and light.

can call upon the "master of tigers" to appear and show them the way home. In Southeast Asia, people believe in a "tiger ancestor" who leads initiates into the jungle where they are symbolically killed and brought back to life as shamans with the ability to shape-shift into tigers. Once in their tiger forms, the shamans can protect their villages with the power of a tiger.

As a fiery, passionate carnivore—a creature who partakes of the flesh—the tiger is associated with drives of all kinds, including those that manifest as insatiable desires and sexual potency. This is why calling someone a "tiger in bed" means he or she has an abundance of sexual energy, whereas asking someone to "tame the tiger" is a request for that person to subdue strong emotions, drives, or impulses. Although both men and women might be described as "tigers," the seductive, energetic tiger is more strongly identified with the feminine principle, in contrast to the lion, who is generally associated with the masculine principle.

INSIGHTS FROM THE TIGER

If ever there were an animal who appears to be completely comfortable in her skin, it's the tiger. She exudes confidence and certainty, such that even when she fails to bring down prey, she appears to casually walk away with an aura of self-assurance, as if she is certain that she'll have plenty of other chances. The tiger's confidence can inspire us to cultivate the same in ourselves, reminding us that there's no point in letting the possibility of failure hold us back.

The tiger also teaches us the value of patience and timing. She will lie still and silent, waiting for her prey for as long as it takes, while calmly keeping her eye on the prize. She reminds us of the virtues of paying attention and biding our time without losing sight of our goals. But when an opportunity arrives, the tiger recognizes it and wastes no time—she pounces, demonstrating that if we want to achieve our goals, we cannot afford to procrastinate or endlessly deliberate.

Tiger moms are doting when cubs are young, but once they have taught them to hunt, cubs must fend for themselves.

Although the tiger calls our attention to our own aggressive natures—to our predatory and even devouring instincts—and asks us if we have these aspects of ourselves under control, she is also reflective of another aspect of our natures. As a creature of majesty, power, and beauty, she invites us to consider our own brilliance and charges us to be brave enough to "burn bright" in our own lives. The tiger has stimulated the imaginations of artists, poets, shamans, and scientists, and she can serve that role for us, too, inspiring us to embrace our fire and live our own fierce grandeur.

**Earth • Endurance • Home • Longevity • Patience • Persistence
Self-protection • Solidity • Survival • Wisdom • World**

TURTLE

THE TURTLE IN NATURE AND CULTURE

One version of a popular story about the origin of the universe begins after a public lecture on cosmology, when an elderly woman approached the presenting scientist to challenge his assertion that the Earth rotates around the sun. When the scientist responded by asking if she had a better theory, she proposed that the Earth rests on the back of a giant turtle. To point out the problem with her assertion, the scientist asked what the turtle is standing on. Not skipping a beat, the woman answered that the first turtle stands on the back of a second, larger turtle. The scientist then asked what this turtle is standing on, to which she confidently replied, "It's turtles all the way down."

The expression "turtles all the way down" is a modern metaphor for the philosophical problem of infinite regress, in which a belief is justified by basing it on another belief that is justified by another belief, and so on. But the idea of a "cosmic turtle" whose body supports the Earth is ancient. In Hindu mythology, the Earth rests on top of four elephants who stand on the shell of a turtle. The turtle also holds up the world in the creation stories of the Lenape and Iroquois peoples of North America. In their myths, the Earth is created by animals piling soil on the back of a turtle. In Chinese tradition, the turtle doesn't merely support the world—she *is* the world. Her domed shell is the heavenly vault, her lower shell is the Earth,

Turtles symbolize tenacity, longevity, and the wisdom that often accompanies a long life.

and her body is the atmosphere in between. The connection between the Earth and the turtle can also be found in Hawaiian culture, in which the words for Earth—*honua*—and turtle—*honu*—are closely related.

The group of animals commonly referred to as turtles are known by three different terms: turtle, tortoise, and terrapin. Although many people use these terms interchangeably, there are differences between the three animals. Turtles are adapted for an aquatic life, with webbed feet or flippers and a more streamlined body. Tortoises are land animals whose bodies are better suited to walking than swimming. Terrapins are smaller turtles who live a dual life on land and in water, especially in swamps and marshes. (Note: the term "turtle" is used here to refer to all three types—turtle, tortoise, and terrapin—unless otherwise specified.)

Turtles are among the oldest and most primitive of reptiles, having evolved nearly three-hundred million years ago. The turtle is the only vertebrate—living or extinct—who has a shell that encases her body.

The turtle's shell is comprised of her spine, ribs, and other bones that are covered in a hard, scale-like skin. With her protective "home" on her back, it's no surprise that the turtle has long symbolized the idea of home—both planetary (as an "Earth bearer") and personal. One of Aesop's fables, "Zeus and the Tortoise," tells us that when Zeus and Hera, king and queen of the gods, were to be married, all of the animals were invited. All attended, except for the tortoise. When asked why she didn't attend, the tortoise said that she preferred staying at home. Offended by her insolence, Zeus decreed that henceforth the tortoise would always carry her home on her back.

Turtles epitomize endurance. Many can go for long periods of time without food or water and, thanks to their shell, can withstand all sorts of physical trauma that would otherwise kill an animal. The turtle's shell is extremely hard, capable of withstanding thousands of pounds of pressure and the assault of most predators. Very few predators can crush a turtle's shell. Even those that try—such as honey badgers, sharks, alligators, and crocodiles—don't always succeed. Golden eagles—who rely on gravity rather than jaw strength—seize a turtle in their talons, fly high, and then drop the turtle onto rocky surfaces, which usually—but not always—cracks open the shell on impact.

Despite the occasional threat from a handful of predators, most of the time the turtle's shell protects her, and so most turtles live long lives, with some tortoises surviving for centuries. As one of the longest lived animals on Earth, turtles symbolize—especially in Asian cultures— tenacity, longevity, and the wisdom that often accompanies a long life. Turtles don't need to move quickly to catch prey, and thanks to their shells, they also do not need to run from predators. Consequently, turtles are slow-moving creatures, which has linked them to patience and persistence. Their slow and steady pace was highlighted in Aesop's fable "The Tortoise and the Hare," in which a sluggish tortoise wins a race against a speedy hare despite the hare having the obvious advantage.

With her protective "home" on her back, it's no surprise that the turtle has long symbolized the idea of home—both planetary (as an "Earth bearer") and personal.

They might be slow, but turtles are nonetheless impressive travelers and navigators. Using the Earth's magnetic field to find their way, sea turtles migrate across oceans every breeding season to return to the same beach where they were born, to lay their eggs. The eggs are left unattended, and when they hatch the baby turtles must find their own way in the world. In Polynesian myth, a World Turtle produced the eggs from which humans were born.

In general, turtles are solitary creatures and their reclusive habits, along with their tendency to withdraw into their shells, connects them to the ideas of isolation, shyness, self-containment, and self-protection.

INSIGHTS FROM THE TURTLE

The turtle is a survivor. She epitomizes endurance and shows us that sometimes the secret to survival—as well as success—is neither fighting nor fleeing, but knowing when to withdraw and wait. The turtle takes adversity in stride and bides her time, waiting patiently and quietly for the right moment to move. Once any threats have passed, she slowly but surely begins to make her way forward again, seemingly unscathed and no less determined.

There is nothing restless about the turtle. She radiates a wise, meditative presence in all she does, which undoubtedly inspired the fabled tortoise who bested the speedy hare. She reminds us that slow and steady wins the race. Having a paced one-step-at-a-time approach to life affords us the opportunity to acquire wisdom and perspective—a peaceful, even-tempered strategy that usually serves us better than acting quickly and recklessly.

Although her shell protects her, the turtle's withdrawal from the world can also symbolize excessive disengagement that can result in isolation and self-imprisonment. So the turtle also cautions us that too much withdrawal shrinks our world, as well as our sense of self. We all need to come out of our shells periodically.

As the mythological "Earth bearer," the turtle carries the "weight of the world" on her back. As such, she connects us to those times when we are saddled with onerous responsibilities and heavy burdens. Although it can feel as if we have no choice about how much weight we carry, sometimes we do. Perhaps burdens could be approached one at a time or shared with others. The biological turtle carries only what she can bear, which prompts us to consider our true carrying capacity and adjust our expectations so that we can better persevere over the long haul.

The turtle cannot crawl out of her shell. She's stuck in it throughout her entire life, just as we are stuck within ourselves throughout our lives. Carrying her home with her wherever she goes, she can inspire us to accept and embrace who we are with patience and grace so that, wherever we go, we are always at home with ourselves.

**Breath • Communication • Empathy • Enormity
Intuition • Journey • Memory • Rebirth • Wisdom**

WHALE

THE WHALE IN NATURE AND CULTURE

The 2011 documentary film *The Whale* tells the story of a young orca who, separated from his family, seeks companionship through interactions with humans. The film's narration compares the encounters with the orca to what it might be like for humans to meet an extraterrestrial, speculating that, "One day we humans may meet an intelligent being from another world. Hollywood tells us this stranger will come flying down in a spaceship, and will look a bit like us. But maybe it won't be like that. Maybe it will be like this."

Whales have always been seen as otherworldly creatures who inspire wonderment and awe. They are warm-blooded mammals who breathe air and give birth to live young, but live almost entirely underwater. One species—Bowhead whales—have the longest lifespan of any mammal on Earth, living up to two hundred years. Another species—Blue whales—are not only the largest animals on the planet at present; they are the largest animals that *ever* lived on Earth, even larger than the most massive of dinosaurs. For these reasons, whales epitomize the idea of enormity and sometimes symbolize the world itself or the foundation that supports it. Islamic tradition, for example, describes the whale as holding up the earth and explains earthquakes as the consequence of the whale shifting under the weight of the world. The

Whales are intelligent, complex, social creatures who demonstrate self-awareness, affection, empathy, grief, and joy.

whale is so big that he was often regarded more as a force of nature than a creature.

In ancient times, whales were seen as leviathans—monsters of the seas—and were linked to the inscrutable mystery of the ocean's depths and its symbolic associations, such as boundlessness, the unconscious, and potentiality. Like other animals who dive deep and then surface to breathe, whales represent movement from the depths to the surface, and the idea of intuition—of unconscious rather than conscious reasoning. Whales don't breathe automatically, as humans and many other animals do, but actively control their breath, usually staying underwater for five to fifteen minutes (and sometimes much longer) before surfacing. Because of their control over their respiration, whales are associated not only with breath, but also with the related ideas of life and spirit.

Whales, along with dolphins and porpoises, are marine mammals known as cetaceans, which are divided into two groups. Baleen whales are the giants of the sea—blue whales, fin whales, humpbacks, and others—who, instead of teeth, have baleen plates that are used to filter small plankton and fish from seawater. Toothed whales—dolphins (including orcas), porpoises, and sperm whales—are predators who use their teeth to catch and eat large prey, such as fish, squid, and sometimes other marine mammals. Both types of whales share a common ancestor, a four-legged creature who lived on land and hunted at the water's edge roughly fifty million years ago. Over time, as these mammals moved deeper into the water, their limbs turned into fins and tails and they became whales, dolphins, and porpoises.

Whales are intelligent, complex, social creatures who demonstrate self-awareness, affection, empathy, grief, and joy. They appear to have culture—behaviors unique to local populations. They form social groups known as pods, help one another, raise their young communally, play together, and maintain long-term friendships. They even come to the aid of other species, such as seals and humans. All these traits and behaviors link whales to empathy and compassion.

Recently, scientists discovered spindle cells in the brains of humpback and other whales. In humans, these cells are associated with emotion, empathy, and social interaction, and have been described as the "cells that make us human." It appears that whales may have three times as many spindle cells as humans do (even after adjusting for brain size), and they may have had this type of cell for twice as long as humans, evolutionarily speaking. So, as it turns out, spindle cells could be thought of as the "cells that make us whale."

Like most animals, whales communicate with one another through body language, but their primary form of communication is sound—whistles, clicks, chirps, moans, rumbles, and other vocalizations—that can travel as far as five hundred miles through water. It appears that whales can even change their vocalizations to match another species. Captive whales have been recorded changing the duration, pitch,

Whales have always been seen as otherworldly creatures and this is especially true for the narwhal, known in ancient times as the "unicorn of the sea." People once believed that narwhal tusks could neutralize poison and cure depression.

and pattern of their calls to match the sounds made by their dolphin cohabitants, possibly suggesting that the whales were trying to "speak dolphin."

The most famous vocalizations made by whales are the hauntingly beautiful songs of humpbacks and other baleen species. Humpback songs are among the most complex forms of animal communication on the planet and consist of repeated phrases, themes, and even rhymes. Sung only by the males, these songs often last for hours. Perhaps the most fascinating characteristic of whale song is that the songs are shared between populations of whales and change over time. No one knows why the songs change or even why whales sing, but numerous theories have been proposed, from attracting mates and socializing to claiming territory and repelling rivals. Another possibility, suggested by philosopher and musician David Rothenberg, is that whales might sing, as we do, for aesthetic reasons.

Whatever the reason, these songs so impressed humankind that we sent their voices into space. In 1977, NASA launched the two Voyager spacecraft that each included a golden record of images, sounds, and greetings intended to represent life on Earth. According to NASA documentation, included on the golden records "were the characteristic 'Hellos' of the humpback whale—another intelligent species from the planet Earth sending greetings to the stars."

Here on Earth, whales travel great distances, with humpbacks covering up to five thousand miles during their seasonal migration. These long journeys played a role in the migration myths of the Maori, the indigenous Polynesian people of New Zealand. According to these legends, whales guided their ancestors on their journeys and showed them where to settle. Similarly, in traditional Hawaiian culture, whales were believed to be manifestations of Kanaloa, the god of the ocean, and to have helped Polynesians discover the Hawaiian Islands.

Perhaps more than anything else, the whale is inextricably linked to the idea of being swallowed. Just as the sea swallows ships, sailors, and swimmers, the whale has long evoked the image of being consumed—both literally and metaphorically. One of the best known of such stories is the Biblical tale of "Jonah and the Whale," in which God called the prophet Jonah to preach to the people of Nineveh. Jonah refused that call, boarded a ship, was thrown overboard, and swallowed by a whale. After Jonah spent three days and nights in the whale's belly, repenting and praying, God commanded the whale to spit him out onto land. Finally willing to answer God's call, Jonah made his way to Nineveh, ready to preach.

Being swallowed by a whale is an example of a "night sea journey," an archetypal motif in which the whale—or other large creature—represents turmoil of such magnitude that it feels as if one is being consumed. Being swallowed signifies the descent into the deepest, darkest phase of inner turmoil. While in the belly of the whale, we are seemingly without resources—our usual strategies, support from family and friends, and sometimes even our faith aren't there for us. But as we struggle to find

Humpback whale songs are among the most complex forms of animal communication on the planet and include rhyming—a compositional technique that was once believed to belong exclusively to humans.

our way in the unfamiliar darkness, we slowly start to figure things out. Emerging from the interior of the whale symbolizes the triumph over the inner struggle and heralds rebirth and new beginnings.

INSIGHTS FROM THE WHALE

When we see whales floating near the surface of the sea, seemingly suspended in space and time, the image evokes an oceanic sensation of undifferentiated consciousness—a breadth of boundlessness in which anything seems possible. But whales also descend to the abyss—to the darkest, deepest, most mysterious parts of the ocean. Offering us a metaphor for exploring our own psychological depths, whales invite us to take the plunge and dive into these deeper inner waters, trusting that when we rise to the surface again, we will have a more expansive, integrated sense of self.

The whale's enormity—and her connection to night sea journey symbolism—suggests the ancient and powerful idea that we can be consumed by something larger than ourselves, struggle with that which consumes us, and emerge stronger, wiser, and whole. Most of us have been "swallowed by a whale" at some point in our lives. Perhaps we were called to undertake a daunting task or make a difficult decision, but we just couldn't rise to the occasion right away. After struggling for a time in the belly of the beast—in the darkness of anxiety and uncertainty—we finally emerged with renewed clarity, strength, and purpose.

By sharing the burdens and joys of life together, the whale pod reminds us of the nurturing and supportive power of community. Their rich repertoire of vocalizations calls our attention to the role communication plays in our relationships. Although we do not know the extent to which their calls, songs, and other utterances mirror our own, their voices nonetheless remind us of the potential to create shared meaning through the signs we use and the songs we sing.

Finally, humans have hunted whales for their meat, blubber, and baleen since prehistoric times, and had hunted them to near extinction. But after scientists discovered and shared the songs of the whale with the world, it radically changed the global perception of these gentle giants of the sea. The discovery forced humankind into taking a collective night sea journey in which we had to face the consequences of commercial whaling—the potential extermination of intelligent, compassionate creatures whose voices we have barely begun to listen to, let alone understand. Fortunately, the growing awareness of whales—their songs, intelligence, and compassion—resulted in a nearly worldwide commercial whaling ban and continues to inspire conservation measures. Although there is still much more work to do to preserve and protect the oceans, whales taught us—and can continue to teach us—to pay attention and listen to the other voices on the planet.

**Community • Courage • Loyalty • Lust • Predation
Stamina • Voracity • Wickedness**

WOLF

THE WOLF IN NATURE AND CULTURE

Once upon a time, the vast forests of Europe were populated with perhaps the most storied creatures of all time: wolves. But they weren't the kind of wolves that most of us imagine today—the charismatic creatures that serve as symbols of wilderness and mascots of the conservation movement. Instead, these wolves were the big, bad, bloodthirsty killers who inspired fairy tales and werewolf legends. They ate grandmothers and little girls. They blew down the houses of innocent little pigs and stuffed their bellies with baby goats. These wolves were, as the fairy tale writer Wilhelm Grimm (of the Brothers Grimm) once said, "the most evil animal of all."

In Western cultures, the wolf has always represented predation, but not simply the ordinary predation of an animal who hunts for a living. The wolf was seen as wicked and deceitful, a view reflected in the Biblical warning of false prophets who were described as "wolves in sheep's clothing." If Christians wandered from Christ, the wolf— a symbol of evil associated with the devil—would be waiting, ready to deceive and devour them. Over time, the wolf came to symbolize destruction of all kinds, which is why, when we narrowly avoid death or ruin, we "keep the wolf from the door."

In myths around the world, the wolf was often cast as a dangerous, crazed, rapacious creature. In Norse mythology, Fenrir, a monstrous

Today, many people see the wolf as a charismatic canid who serves as a symbol of wilderness, whereas in the not-so-distant past, he was seen as a creature closer to the big, bad, bloodthirsty wolves of fairy tales and legends.

wolf, is kept chained to a rock far beneath the earth, with a sword wedged between his jaws to keep him from devouring the world. The depiction of the wolf's insatiable hunger can also be found in the mythical wolfish creature of European medieval legend, the werewolf. Werewolves walked the world as humans, but when the moon was full, they shape-shifted into their bestial forms and ate men, women, children, livestock—everything in their paths—because they couldn't control their appetites. This view of the wolf's voracity persists even today, which is why someone eating ravenously might be described as "wolfing it down."

The words for wolf in several European languages are yet another reflection of the pejorative attributes projected onto the maligned canid. For instance, the Swedish and Norwegian term *varg* means wolf, as well as wicked person; and the Old High German term for wolf, *Warg*, was also used for an outlaw, murderer, or evil spirit. Clearly, wolves frightened people and triggered primal fears of being devoured, which is understandable given that wolves are predators capable of bringing down prey much bigger and faster than humans. However, by demonizing the wolf—by interpreting his predation not simply as a way to stay alive, but as an expression of moral corruption—people were doing more than venting fear. They were projecting onto the wolf those impulses that humankind often denies in itself, such as savagery, thievery, and deceitfulness.

European cultures even projected lechery onto wolves. Wolves were seen as lusting for flesh in *every* way—not just as food—which is why the Latin word for a female wolf—*lupa*—also meant whore. Similarly, in seventeenth-century France, when a girl lost her virginity, one might have said that "*elle avoit vû le loup*," which translates as "she had seen the wolf." A more recent example of the wolf's association with saliciousness is the "wolf whistle," a two-note whistle used to express sexual interest, usually by a man for a woman.

The wolf usually howls to call or reassure pack mates, or to establish territory with rival wolves.

Despite their reputations as malicious predators, there are also myths and legends of benevolent, nurturing wolves, such as the Roman myth of Romulus and Remus, in which a female wolf nursed abandoned human infants who grew up to found the city of Rome. There also are scattered legends throughout history of children being raised by wolves—an idea popularized in stories such as Rudyard Kipling's *Jungle Book*, in which the character Mowgli is raised by a wolf pack. These tales—in which the child must eventually shed his wolfish ways in order to return to human society—inspired the phrase "raised by wolves," used to describe someone who is uncivilized and socially inept.

As for the real wolf, he is a large, intelligent, gregarious canid whose closest cousin is the domestic dog. In fact, dogs and wolves are so closely related that DNA analysis cannot distinguish one from the other. Wolves once lived throughout North America and Eurasia, but as the result of centuries-long extermination efforts, the two remaining species, gray wolves and red wolves—and the various subspecies related to each of them—have been largely eliminated from most of Europe and the United States.

As the most socially evolved of all the canids, wolves prioritize relationships. They often mate for life and live in packs consisting of a mated pair, their offspring, and sometimes unrelated juveniles. Members of a pack bond, bicker, and reconcile, just like any human family does. They play together—not only as pups, but throughout their lives—wrestling, sparring, ambushing, and chasing one another, often taking turns to reverse roles. Like dogs, they play with sticks, tossing them into the air, leaping up to catch them, then dropping and pouncing on them as if they were prey. Wolves are very affectionate, constantly nuzzling, licking, grooming, and leaning against one another. They raise pups communally, with aunts, uncles, and older siblings helping out with feeding, guarding, and playing with the pups. They also care for older and injured pack members who cannot hunt

Rudyard Kipling said it best: "As the creeper that girdles the tree trunk, the law runneth forward and back; / For the strength of the pack is the wolf, and the strength of the wolf is the pack."

by sharing food with them. As a result of these behaviors, wolves are associated with community, loyalty, and fidelity.

Given that relationships are so important to wolves, it's not surprising that they have evolved a complex communication system involving vocalizations, gestures, facial expressions, and scent marking. The wolf's best known vocalization—the howl—is most often used to call or reassure pack mates, or to establish territory with rival wolves. To human ears, the howl of the wolf symbolizes the call of the wild and the yearning for connection with others.

The wolf in nature, like the wolf in folklore, hunts and kills in order to eat. But wildlife biologists—as well as many indigenous hunting cultures—see wolves not as crazed killers, but as strategic, cooperative hunters. Wolves often coordinate their hunts, with each wolf contributing however they can to bring down prey and thereby help provide for their pack. As for the wolf's reputation for voracity, it originates in their feast or famine lifestyle. When food is available, wolves devour it in large quantities. During times of scarcity, they might go a week or longer without eating while traveling as much as ten hours a day in search of prey.

The Pawnee, a Central Plains Indian tribe, were so impressed by the way wolves hunted and worked together that they emulated them in their warfare and hunting strategies. Known as the Wolf People, other tribes believed that the Pawnees had a wolfish acuity of vision and hearing well beyond that of ordinary humans. When sneaking into enemy territory or herds of bison, they would often wear wolf pelts, drop onto their hands and knees, and move like wolves. To the Pawnee, the wolf not only epitomized courage, stamina, and skill in the hunt, but also devotion to their pack, which they saw as analogous to loyalty to one's tribe. The Pawnee called the brightest star in the sky—known to Western astronomers as Sirius—the Wolf Star, and their name for the Milky Way was the Wolf Road, reflecting the high regard they had for their canid mentors.

INSIGHTS FROM THE WOLF

In his poem "The Law for the Wolves" (also known as "The Law of the Jungle"), Rudyard Kipling presents a code of behavior for wolves, summed up by the couplet, "As the creeper that girdles the tree trunk, the law runneth forward and back; / For the strength of the pack is the wolf, and the strength of the wolf is the pack." The wolf pack shows us that, for the most part, life is easier and more satisfying if one belongs to a community. Times of hardship are better endured and times of triumph are more meaningful when we share them with those close to us.

Despite wolves' loyalty and devotion to one another, pack members will take time away from the group to scout out prey, reminding us that even when we belong to a pack, we still need to wander off on our own occasionally. As a pathfinder who travels great distances every day, the wolf can inspire us to explore new territory and blaze new trails.

Humankind's polarized views of the wolf invite us to think about how our perceptions of wolves—and all animals—shape the way we treat them.

In doing so, we are likely to return with new ideas and resources that will benefit our circle of close relationships, as well as ourselves. In this way—as both a pack member and a trail blazer—the wolf teaches us to seek a balance between our individual needs and those of our group.

The essence of a wolf's life, like our own, is love, work, and play, reminding us that all three are equally important to living a full and balanced life. The wolf also shows us that effective communication is essential in each of these arenas. The wolf expresses affection, avoids conflicts, conveys hunting strategies, mentors the young, and offers invitations to play using a rich repertoire of non-verbal signals. Like wolves, we, too, rely on body language. Eye contact, tone of voice, gesture, posture, and other nonverbal cues send strong messages that enhance or undermine what we are trying to say. The wolf can inspire us to pay attention to the nonverbal cues we give and receive, and to remember that an arm around a shoulder or pat on the back often make the strongest impression.

More than any other animal, the wolf evokes our fear of the wild because he represents the fierce power of nature, as well as our uneasiness with the untamed, beastly parts of ourselves. And yet, the wolf also evokes our love of the wild because he connects us to the beauty and mystery in nature, as well as to our longing to connect with our own wild hearts. These two polarized views of the wolf call to mind the old Cherokee legend known as "The Wolves Within." In the legend, a grandfather explains the concept of inner conflict to his grandson, using a metaphor of a person who has two wolves fighting for dominance within. One wolf is filled with darkness, fear, anger, greed, regret, resentment, dishonesty, doubt, and hate; the other is filled with light, courage, peace, generosity, hope, contentment, truth, faith, and love. When the grandson asks which wolf will win the fight, the grandfather answers, "The one you feed."

KEYWORD INDEX

Abundance: Bison, Mouse and Rat, Pig

Adaptability: Coyote, Fox, Lizard, Mouse and Rat, Octopus

Aggression: Bear

Agility: Dragonfly, Goat, Otter

Altruism: Dolphin

Ambiguity: Bat

Ambition: Shark

Balance: Crane, Goat, Seal and Sea Lion

Beauty: Peacock, Swan

Breath: Dolphin, Whale

Capriciousness: Goat

Change: Nautilus, Octopus

Changeability: Crab

Chaos: Crocodile

Cleverness: Coyote, Rabbit and Hare

Communication: Dolphin, Whale

Community: Ant, Dog, Dolphin, Elephant, Honey Bee, Parrot, Sheep, Wolf

Compassion: Elephant, Goose

Confidence: Badger, Cat, Peacock, Tiger

Cooperation: Ant, Badger, Beaver, Honey Bee, Lion

Courage: Badger, Hummingbird, Otter, Wolf

Creativity: Bear, Beaver, Coyote, Crocodile, Honey Bee, Raven and Crow, Snake, Spider

Cunning: Fox, Mouse and Rat

Curiosity: Otter

Cycles: Crab, Crane, Goose

Darkness: Bat, Owl

Death: Bat

Deception: Crocodile, Snake, Spider

Depth: Nautilus, Octopus

Desire: Horse

Destruction: Snake, Spider

Determination: Badger, Beetle

Devotion: Dog, Goose

Dexterity: Ape and Monkey, Otter

Diligence: Ant, Beaver

Display: Peacock

Divinity: Eagle

Domesticity: Badger, Beaver, Duck

Dominion: Eagle, Lion

Drive: Tiger

Duality: Eagle, Snake

Earth: Turtle

Efficiency: Shark

Elegance: Swan

Elusiveness: Deer, Jaguar, Mouse and Rat, Octopus, Tiger

Emotional Comfort: Duck

Empathy: Whale

Empowerment: Horse

Endurance: Badger, Camel, Crocodile, Turtle

Energy: Horse, Hummingbird

Enormity: Whale

Enthusiasm: Beaver

Evasion: Badger

Evil: Snake

Evolution: Frog, Nautilus

Excess: Pig

Expression: Lizard

Exuberance: Goat

Fate: Spider

Fatherhood: Seahorse

Fear: Snake, Spider

Fellowship: Goose

Ferocity: Crocodile, Hawk and Falcon

Fertility: Bison, Crocodile, Frog, Goat, Goose, Honey Bee, Mouse and Rat, Pig, Rabbit and Hare, Sheep, Snake

Fidelity: Crane, Goose, Parrot, Swan

Flexibility: Crane

Flow: Nautilus

Independence: Cat, Goat

Industry: Ant, Beaver

Infestation: Mouse and Rat

Ingenuity: Fox, Rabbit and Hare

Inner Resources: Camel

Innocence: Sheep

Inspiration: Swan

Instinct: Ape and Monkey, Crocodile, Horse, Shark

Intelligence: Parrot

Interconnectedness: Bison

Intuition: Horse, Seal and Sea Lion, Whale

Invisibility: Owl

Involution: Nautilus

Journey: Camel, Crane, Goose, Whale

Largeness: Elephant

Life Force: Bison, Snake

Light: Dragonfly, Eagle, Raven and Crow, Swan

Liminality: Cat

Longevity: Crane, Turtle

Love: Parrot, Seahorse, Swan

Loyalty: Dog, Elephant, Goose, Lion, Seahorse, Wolf

Luck: Rabbit and Hare, Seahorse

Lust: Wolf

Magic: Dragonfly, Hummingbird, Owl, Rabbit and Hare, Seahorse

Majesty: Tiger

Mastery: Tiger

Productivity: Honey Bee

Prophecy: Owl

Protection: Dog, Nautilus

Prudence: Ant

Purity: Sheep, Swan

Reaction: Frog

Rebirth: Bat, Bear, Beetle, Crab, Frog, Snake, Whale

Regeneration: Crab, Deer

Renewal: Beetle, Crab, Frog

Resilience: Cat

Resoluteness: Elephant

Resourcefulness: Coyote

Restlessness: Shark

Reversal: Seahorse

Reward: Honey Bee

Sacrifice: Sheep

Secrecy: Owl

Seduction: Deer, Fox

Self-containment: Cat

Self-development: Beetle

Self-protection: Crab, Turtle

Self-reliance: Bear

Selflessness: Ant

Sensitivity: Deer, Duck, Frog, Lizard, Rabbit and Hare

Service: Camel

Sexuality: Snake

Transformation: Beetle, Butterfly, Coyote, Dragonfly, Frog, Jaguar, Raven and Crow, Seal and Sea Lion, Swan

Transition: Bat

Trickster: Ape and Monkey, Coyote, Fox, Rabbit and Hare, Raven and Crow

Trust: Horse, Otter

Uncleanliness: Pig

Unconscious: Seal and Sea Lion

Vanity: Peacock

Vision: Eagle, Hawk and Falcon

Vitality: Goat, Hummingbird

Vivacity: Otter

Voracity: Pig, Shark, Wolf

Vulnerability: Sheep

Warfare: Hawk and Falcon

Watchfulness: Hawk and Falcon

Water: Frog, Snake

Wickedness: Wolf

Wildness: Cat, Hawk and Falcon

Wisdom: Bear, Elephant, Owl, Raven and Crow, Snake, Spider, Turtle, Whale

Withdrawal: Nautilus

Wonder: Seahorse

World: Turtle

SELECTED BIBLIOGRAPHY

This selected bibliography does not attempt to list every source that has influenced my thinking in writing this book. Having written two prior books as well as numerous articles on a wide range of topics related to animals and the human-animal relationship, it would be impossible to honor everything that has contributed to my perspective. Given the nature of this book, I have chosen to include those sources that I believe would be most helpful for readers seeking to pursue these ideas further.

Adams, Christina. *Camel Crazy: A Quest for Miracles in the Mysterious World of Camels.* Novato, CA: New World Library, 2019.

Aldhouse-Green, Miranda Jane. *Animals in Celtic Life and Myth.* London: Routledge, 1992.

Allen, Daniel. *Otter.* London: Reaktion Books, 2010.

Andrews, Tamra. *Dictionary of Nature Myths: Legends of the Earth, Sea, and Sky.* Oxford: Oxford University Press, 2000.

Archive for Research in Archetypal Symbolism (ARAS). Martin, Kathleen, ed. *The Book of Symbols.* Köln: Taschen, 2010.

Armstrong, Philip. *Sheep.* London: Reaktion Books, 2016. Kindle.

Arnold, Jennifer. *Through a Dog's Eyes.* New York: Random House Publishing Group, 2010.

Balcombe, Jonathan. *Second Nature: The Inner Lives of Animals.* New York: St. Martin's Press, 2010.

————. *What a Fish Knows: The Inner Lives of Our Underwater Cousins.* New York: Scientific American / Farrar, Straus and Giroux, 2016.

Barkham, Patrick. *Badgerlands: The Twilight World of Britain's Most Enigmatic Animal,* rev. ed. London: Granta Books, 2014.

Bearzi, Maddalena and Craig B. Stanford. *Beautiful Minds: The Parallel Lives of Great Apes and Dolphins.* Cambridge, MA: Harvard University Press, 2008.

Bekoff, Marc. *The Emotional Lives of Animals.* Novato, CA: New World Library, 2007.

Bekoff, Marc, and Jessica Pierce. *Wild Justice: The Moral Lives of Animals.* Chicago: The University of Chicago Press, 2009.

Bettelheim, Bruno. *The Uses of Enchantment: The Meaning and Importance of Fairy Tales.* New York: Vintage Books, 2010.

Bieder, Robert E. *Bear.* London: Reaktion Books, 2005.

Bleakley, Alan. *The Animalizing Imagination: Totemism, Textuality and Ecocriticism.* New York: St. Martin's Press, 2000.

Brown, Joseph Epes. *Animals of the Soul: Sacred Animals of the Oglala Sioux,* rev. ed. Rockport, MA: Element Books, 1997.

Brunner, Bernd. *Bears: A Brief History.* New Haven, CT: Yale University Press, 2007.

Burgoyne, John, and Jennifer Ackerman. *The Genius of Birds.* New York: Penguin Press, 2016.

Campbell, Joseph. *The Hero with a Thousand Faces.* Novato, CA: New World Library, 2008.

Carnell, Simon. *Hare.* London: Reaktion Books, 2010.

Carroll, Georgie. *Mouse*. London: Reaktion Books, 2015. Kindle.

Carter, Paul. *Parrot*. London: Reaktion Books, 2006.

Caspari, Elizabeth, with Ken Robbins. *Animal Life In Nature, Myth and Dreams*. Wilmette, IL: Chiron Publications, 2003.

Chevalier, Jean, and Alain Gheerbrant. *The Penguin Dictionary of Symbols*, 2nd ed. London: Penguin Books, 1996.

Cirlot, Juan Eduardo. *A Dictionary of Symbols*, 2nd ed. New York: Philosophical Library, 1983.

Coates, Peter. *Salmon*. London: Reaktion Books, 2006.

Cooper, Simon. *The Otters' Tale*. London: HarperCollins Publishers, 2017.

Cox, Lynne. *Grayson*. Orlando: Harcourt, 2008.

Crawford, Dean. *Shark*. London: Reaktion Books, 2008.

de Rijke, Victoria. *Duck*. London: Reaktion Books, 2008.

de Waal, Frans. *Are We Smart Enough to Know How Smart Animals Are?* New York: W. W. Norton, 2017.

———. *Mama's Last Hug: Animal Emotions and What They Tell Us About Ourselves*. New York: W. W. Norton, 2019.

Dickenson, Victoria. *Seal*. London: Reaktion Books, 2016.

Dobie, James Frank. *The Voice of the Coyote*. Lincoln: University of Nebraska Press, 2006.

Dorson, Richard Mercer, Yoshie Noguchi, and Wolfgang Laade. *Folk Legends of Japan*. Rutland, VT: C.E. Tuttle, 1961.

Eastman, K.P., and G.I. Omura. *Folklore Around the World: An Annotated Bibliography of Folk Literature*. Honolulu: School of Library and Information Studies, University of Hawaii at Manoa, 1994.

Evans, Ivor H., ed. *Brewer's Dictionary of Phrase and Fable*, Centenary ed., rev. New York: Harper & Row, 1981.

Fletcher, John. *Deer*. London: Reaktion Books, 2013.

Flores, Dan. *Coyote America: A Natural and Supernatural History*. New York: Basic Books, 2016. Kindle.

Foster, Charles. *Being a Beast: Adventures Across the Species Divide*. New York: Picador, 2017.

Fouts, Roger, with Stephen Mills. *Next of Kin: My Conversations with Chimpanzees*. New York: HarperCollins, 2003. First published in 1997 by William Morrow.

Franklin, Jon. *The Wolf in the Parlor: How the Dog Came to Share Your Brain*. New York: Henry Holt and Company, 2009.

Gibson, Graeme. *The Bedside Book of Beasts: A Wildlife Miscellany*. New York: Nan A. Talese / Doubleday, 2009.

Giggs, Rebecca. *Fathoms: The World in the Whale*. New York: Simon & Schuster, 2020.

Godfrey-Smith, Peter. *Other Minds: The Octopus and the Evolution of Intelligent Life*. New York: Farrar, Straus and Giroux, 2016.

Green, Susie. *Tiger*. London: Reaktion Books, 2006.

Hannah, Barbara. *The Archetypal Symbolism of Animals: Lectures Given at the C. G. Jung Institute, Zurich, 1954-1958*. Wilmette, IL: Chiron Publications, 2006.

Harrod, Howard L. *The Animals Came Dancing: Native American Sacred Ecology and Animal Kinship*. Tucson: University of Arizona Press, 2000.

Heinrich, Bernd. *The Geese of Beaver Bog*. New York: HarperCollins, 2004.

———. *Mind of the Raven: Investigations and Adventures with Wolf-Birds*. New York: HarperCollins, 1999.

Hillman, James. *Animal Presences: Uniform Edition of the Writings of James Hillman*. Putnam, CT: Spring Publications, 2008.

Hyde, Lewis. *Trickster Makes This World: Mischief, Myth, and Art*. New York: Farrar, Straus and Giroux, 2010.

Irwin, Robert. *Camel*. London: Reaktion Books, 2010.

Jackson, Christine E. *Peacock*. London: Reaktion Books, 2006.

Jackson, Deirdre. *Lion*. London: Reaktion Books, 2010.

Justice, Daniel Heath. *Badger*. London: Reaktion Books, 2014.

Kilham, Benjamin. *In the Company of Bears: What Black Bears Have Taught Me about Intelligence and Intuition*. White River Junction, VT: Chelsea Green Publishing, 2014.

King, Barbara J. *Being With Animals: Why We Are Obsessed with the Furry, Scaly, Feathered Creatures Who Populate Our World*. New York: Doubleday, 2010.

Layard, John. *The Lady of the Hare: Being a Study of the Healing Power of Dreams*. London: Routledge, 2011.

Leeming, Margaret, and David Leeming. *A Dictionary of Creation Myths*. Oxford: Oxford University Press, 1994.

Lévi-Strauss, Claude. *Totemism*. Boston, MA: Beacon Press, 1967.

Lopez, Barry H. *Arctic Dreams: Imagination and Desire in a Northern Landscape*. New York: Vintage Books, 2001.

———. *Giving Birth to Thunder, Sleeping with His Daughter: Coyote Builds North America*. New York: HarperCollins, 1990.

———. *Of Wolves and Men*. New York: Charles Scribner's Sons, 1978.

Lyman, Darryl. *The Animal Things We Say.* New York: Jonathan David Publishers, 1983.

Macdonald, Helen. *Falcon.* London: Reaktion Books, 2016.

———. *H is for Hawk.* New York: Grove Press, 2016.

———. *Vesper Flights: New and Collected Essays.* New York: Grove Press, 2020.

Malamud, Randy. *Poetic Animals and Animal Souls.* New York: Palgrave Macmillan, 2003.

Marvin, Garry. *Wolf.* London: Reaktion Books, 2012.

Marzluff, Colleen, and John M. Marzluff. *Dog Days, Raven Nights.* New Haven, CT: Yale University Press, 2011.

Marzluff, John M., and Tony Angell. *In the Company of Crows and Ravens.* New Haven, CT: Yale University Press, 2008.

———. *Gifts of the Crow: How Perception, Emotion, and Thought Allow Smart Birds to Behave Like Humans.* New York: Atria Books, 2013.

Masson, Jeffrey Moussaieff. *Dogs Never Lie About Love: Reflections on the Emotional World of Dogs.* New York: Crown Publishers, 1997.

———. *The Pig Who Sang to the Moon: The Emotional World of Farm Animals.* New York: Ballantine Books, 2003.

Masson, Jeffrey Moussaieff, and Susan McCarthy. *When Elephants Weep.* New York: Delacorte Press, 1995.

McHugh, Susan. *Dog.* London: Reaktion Books, 2004.

McLean, Margot, and James Hillman. *Dream Animals.* San Francisco: Chronicle Books, 1997.

Michalski, Katarzyna, and Sergiusz Michalski. *Spider.* London: Reaktion Books, 2010. Kindle.

Mizelle, Brett. *Pig*. London: Reaktion Books, 2011. Kindle.

Montgomery, Sy. *The Good Good Pig: The Extraordinary Life of Christopher Hogwood*. New York: Ballantine Books, 2006. Kindle.

———. *The Soul of an Octopus: A Surprising Exploration Into the Wonder of Consciousness*. New York: Atria Books, 2016.

———. *Spell of the Tiger: The Man-eaters of Sundarbans*. White River Junction, VT: Chelsea Green Publishing, 2008.

Montgomery, Sy, and Rebecca Green. *How to Be a Good Creature: A Memoir in Thirteen Animals*. Boston, MA: Houghton Mifflin Harcourt, 2018.

Moon, Beverly, ed. *An Encyclopedia of Archetypal Symbolism*. Boston, MA: Shambhala, 1991.

Morris, Desmond. *Owl*. London: Reaktion Books, 2009.

Moyers, Bill, Joseph Campbell, and Betty S. Flowers. *The Power of Myth*. New York: Anchor Books, 1991.

O'Brien, Stacey. *Wesley the Owl: The Remarkable Love Story of an Owl and His Girl*. New York: Free Press, 2008.

Pepperberg, Irene. *Alex & Me: How a Scientist and a Parrot Discovered a Hidden World of Animal Intelligence—and Formed a Deep Bond in the Process*. New York: HarperCollins, 2009.

Poliquin, Rachel. *Beaver*. London: Reaktion Books, 2015. Kindle.

Preston, Claire. *Bee*. London: Reaktion Books, 2006.

Quammen, David. *Monster of God: The Man-Eating Predator in the Jungles of History and the Mind*. New York: W. W. Norton, 2004.

Recio, Belinda. *Inside Animal Hearts and Minds: Bears That Count, Goats that Surf, and Other True Stories of Animal Intelligence and Emotion*. New York: Skyhorse Publishing, 2017.

————. *When Animals Rescue: Amazing True Stories about Heroic and Helpful Creatures*. New York: Skyhorse Publishing, 2020.

Rockwell, David. *Giving Voice to Bear: North American Indian Myths, Rituals, and Images of the Bear*. Lanham, MD: Roberts Rinehart Publishers, 2003.

Rogers, Janine. *Eagle*. London: Reaktion Books, 2014.

Rogers, Katharine M. *Cat*. London: Reaktion Books, 2006.

Roman, Joe. *Whale*. London: Reaktion Books, 2006.

Rothenberg, David. *Thousand Mile Song: Whale Music in a Sea of Sound*. New York: Basic Books, 2008.

————. *Survival of the Beautiful: Art, Science, and Evolution*. New York: Bloomsbury Press, 2011.

Roud, Stephen, and Jacqueline Simpson. *A Dictionary of English Folklore*. Oxford: Oxford University Press, 2003.

Russack, Neil. *Animal Guides in Life, Myth and Dreams: An Analyst's Notebook*. Toronto: Inner City Books, 2002.

Safina, Carl. *Becoming Wild: How Animal Cultures Raise Families, Create Beauty, and Achieve Peace*. New York: Henry Holt, 2020.

Saunders, Nicholas J. *The Cult of the Cat*. London: Thames and Hudson, 1991.

Sax, Boria. *The Mythical Zoo: An Encyclopedia of Animals in World Myth, Legend, and Literature*. Santa Barbara: ABC-CLIO, 2001.

————. *Crow*. London: Reaktion Books, 2004.

————. *Lizard*. London: Reaktion Books, 2017. Kindle.

Scales, Helen. *Poseidon's Steed: The Story of Seahorses, from Myth to Reality*. New York: Penguin Group (USA), 2009.

Schweid, Richard. *Octopus.* London: Reaktion Books, 2013.

Shepard, Paul. *The Others: How Animals Made Us Human.* Washington, D.C.: Island Press, 1997.

———. *Traces of an Omnivore.* Washington, D.C.: Island Press, 1996.

———. *Thinking Animals: Animals and the Development of Human Intelligence.* Athens: University of Georgia Press, 2011.

Shepard, Paul, and Barry Sanders. *The Sacred Paw: The Bear in Nature, Myth, and Literature.* New York: Viking Penguin, 1985.

Sleigh, Charlotte. *Ant.* London: Reaktion Books, 2003. Kindle.

———. *Frog.* London: Reaktion Books, 2012. Kindle.

Sorenson, John. *Ape.* London: Reaktion Books, 2009.

Carter, Nancy, ed. *Spring Journal Vol. 83, Summer 2010: Minding the Animal Psyche.* New Orleans: Spring Journal, 2010.

Stutesman, Drake. *Snake.* London: Reaktion Books, 2005.

Taylor, Marianne. *The Way of the Hare.* London: Bloomsbury Wildlife, 2017.

Velten, Hannah. *Cow.* London: Reaktion Books, 2007.

Vesey-Fitzgerald, B. *A Country Chronicle.* London: Chapman & Hall, 1942.

Walker, Elaine. *Horse.* London: Reaktion Books, 2008.

Wallen, Martin. *Fox.* London: Reaktion Books, 2006.

Werness, Hope B. *The Continuum Encyclopedia of Animal Symbolism in Art.* New York: Continuum, 2006.

Wilmore, Sylvia Bruce. *Swans of the World.* New York: Taplinger Publishing Company, 1979.

Wylie, Dan. *Elephant.* London: Reaktion Books, 2009.

Young, Peter. *Tortoise.* London: Reaktion Books, 2004.

———. *Swan.* London: Reaktion Books, 2008.

Zipes, Jack. *The Oxford Companion to Fairy Tales.* Oxford: Oxford University Press, 2000.